NUCLEIC ACID–METAL ION INTERACTIONS

METAL IONS IN BIOLOGY

EDITOR: **Thomas G. Spiro**
Department of Chemistry
Princeton University, Princeton, New Jersey 08540

VOLUME 1   *Nucleic Acid—Metal Ion Interactions*

In Preparation
Volume 2 Metal Ion Activation of Dioxygen
Volume 3 Copper Proteins
Volume 4 Iron—Sulphur Proteins

Nucleic Acid–Metal Ion Interactions

Edited by

T:HOMAS G. SPIRO
Princeton University

A WILEY-INTERSCIENCE PUBLICATION

JOHN WILEY & SONS, New York · Chichester · Brisbane · Toronto

Library of Congress Cataloging in Publication Data:

Main entry under title:
Nucleic acid–metal ion interactions.

  (Metal ions in biology; v. 1)
  Includes index.
  1. Nucleic acids.   2. Metals—Physiological effect.
I. Spiro, Thomas G., 1935–   II. Series.

QP620.N78   574.8′732   79-13808
ISBN 0-471-04399-0

Printed in the United States of America

10 9 8 7 6 5 4 3 2 1

Series Preface

Metal ions are essential to life as we know it. This fact has long been recognized, and the list of essential "trace elements" has grown steadily over the years, as has the list of biological functions in which metals are known to be involved. Only recently have we begun to understand the structural chemistry operating at the biological sites where metal ions are found. This has come about largely through the application of powerful physical and chemical structure probes, particularly X-ray crystallography, to purified metalloproteins. From such studies we have learned that nature has evolved highly sophisticated ways of controlling the relatively flexible stereochemistry of metal ions. In one case after another, the structure and reactivity of a metalloprotein active site has turned out to be different from anything previously encountered in simple compounds of the metals. Indeed, many a reasonable inference about active site structure, based on the known properties of metal complexes in solution, have turned out to be erroneous. These surprises have inspired inorganic chemists to expand their vision of metal ion reactivity. The biological studies have spurred much fruitful synthetic and mechanistic work in inorganic chemistry, aimed at elucidating the means whereby nature achieves its stereochemical ends. The terra incognita of the biochemical functions of metal ions has become familiar territory to an increasing number of inorganic chemists and biochemists, and several of the more imposing mountains have been scaled. Vast stretches remain uncharted, and the field is alive with a sense of both accomplishment and new opportunities.

The purpose of this series is to convey some of this excitement, as well as the emerging intellectual shape of the field, to a wide audience of nonspecialists. Individual volumes will cover topics that are current and exciting—the recently scaled mountains that are still under active exploration. The chapters are not intended to be exhaustive reviews of the subject matter. Rather, they are intended to be readable accounts of the insights and directions that are emerging in active new areas of research. Volumes will appear on an occasional basis as progress in the field dictates.

THOMAS G. SPIRO

Princeton, New Jersey

Preface

It is appropriate that this series begin with a consideration of the interactions of metal ions with nucleic acids in view of the dramatic strides currently made in our understanding of the structural basis of nucleic acid function and the important roles played by metal ions. Much excitement has been generated by the discovery and therapeutic application of the anticancer properties of *cis*-dichlorodiammineplatinum(II) and related metal complexes. It is fitting that the volume begin with an overview of this development by its pioneer, Barnett Rosenberg. In Chapter 2, Jacqueline K. Barton and Stephen J. Lippard give a thoughtful discussion of the possible modes of action of the platinum anticancer agents and also of the uses of heavy metal complexes in probing the structure of nucleic acids. Chapter 3, by Lawrence A. Loeb and Richard A. Zakour, takes up the subject of metals as mutagens and carcinogens and summarizes the interesting work of the Loeb group on errors in DNA replication induced by metal ions. Chapter 4, by Martha M. Teeter, Gary J. Quigley, and Alexander Rich, is a lucid account of metal ion interactions with transfer RNA, the one nucleic acid for which there are high-resolution three-dimensional crystal structures. In Chapter 5, Luigi G. Marzilli, Thomas J. Kistenmacher, and Gunther L. Eichhorn provide a systematic discussion of the modes whereby metal ions may interact with the constituents of nucleic acids, as deduced from a variety of physical studies. These chapters provide the reader with much food for thought in this rich field and with a useful orientation to several of the avenues of current research.

THOMAS G. SPIRO

Princeton, New Jersey
May 1979

vii

Contents

NUCLEIC ACID–METAL ION INTERACTIONS

CHAPTER **1**

Platinum Complexes for the Treatment of Cancer

BARNETT ROSENBERG

Michigan State University, East Lansing

CONTENTS

Cancer research has escalated enormously in this past decade. Despite this, it is still a rare occurrence for a new class of anticancer drugs to surface. In the random screening of chemicals, only about 1 in 10,000 shows significant activity to warrant further tests. It is all the more interesting when a new class of drugs emerges from an unexpected area of chemistry. Therefore, it is not surprising that the report of anticancer activity of platinum coordination complexes from this laboratory in 1969 evoked strong but mixed reactions. There was joy amongst inorganic chemists, who, for too long, have been largely excluded from medical research; and there was skepticism among medical scientists, who are deeply conditioned to consider heavy metal compounds as poisons. Now that the first platinum drug, cis-dichlorodiammineplatinum(II), has proved to be of value in treating many human cancers, it may not be inappropriate to describe briefly the history of its discovery and the subsequent tortuous developments leading to its clinical use.

1 EARLY HISTORY

The story begins in 1961, when I left the Physics Department at New York University to help found the Biophysics Department at Michigan State University. With this change of departments there came an obligation to orient my research more toward biology. In my earlier reading I had been fascinated by the microphotographs of the mitotic figures in cells in process of division. They called to a physicist's mind nothing so much as the shape of an electric or magnetic dipole field, the kind one sees with iron filings over a bar magnet.

If such a dipole may be involved in cell division—as some had earlier speculated—then by tickling the dipole with electromagnetic radiation of a resonant frequency, or a subharmonic, to avoid the radiofrequency heating of cells, it may absorb some energy which may or may not be detrimental to the cell. Admittedly this is an overburden of 'mays', but I was intrigued by the idea of an experimental test. Having no competence in biology—few physicists have—L. VanCamp joined the laboratory to do the test. We set up a continuous culture apparatus for the cells, but included in the growth chamber a set of platinum electrodes. Platinum of course, is known to be quite inert in a biologic environment.

The electrodes were powered by an audio amplifier whose input frequency was set by an audio oscillator. The impedance of the chamber, 6 Ω, was perfectly matched to the output impedance of the amplifier. To test the proper functioning of the apparatus before putting in mammalian cells, we used the common bacterium Escherichia coli. These, and pro-karyotic cells generally, do not show mitotic figures in division. After the bacterial population reached a steady state, the electric field was turned on. The density of bacteria started to decline, and we were in danger of having an aseptic chamber. When the field was turned off, the density returned to normal after a few hours.

A rather striking effect, but how striking we did not realize until we examined the bacterial cells in the effluent of the chamber. The bacterial rods normally look like the picture in Fig 1a, rods about 2 to 5 μm long, with a 1 μm diameter. After an exposure to the electric field they appeared

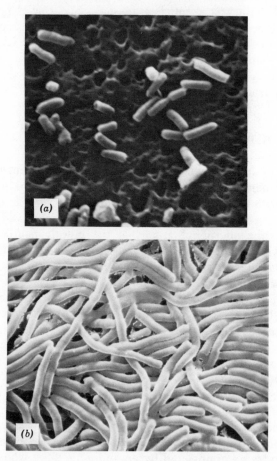

Figure 1 (a) Scanning electron microphotograph of normal *E. coli* (Gram-negative rods). (b) Scanning electron microphotograph of *E. coli* grown in medium containing a few parts per million of *cis*-dichlorodiammineplatinum(II). Same magnification in both pictures. The platinum drug has inhibited cell division, but not growth, leading to long filaments. These pictures were taken by D. Beck of Bowling Green University.

as in Fig 1b; long filaments, up to 300 times the usual length. Now this required an explanation. The effect was not due to a direct action of the electric field on the bacterial cell but rather to electrolysis products from the platinum electrodes.

We now brought our chemist, T. Krigas, in to isolate and identify these products. He clearly identified it as a platinum containing compound, probably ammonium chloroplatinate $(NH_4)_2[PtCl_6]$. We were somewhat nonplussed, however, when addition of this compound at the detected concentration to bacterial cells in test tube cultures led, not to filamentation, but to bacteriocidal activity. Many experiments later we found that a solution of this compound, after standing on our laboratory shelf for a few weeks, was able to produce a small amount of short filaments.

Some quick studies showed that light was the necessary agent for the change, and we were now deep into the photochemistry of platinum. In retrospect this was not surprising. Platinum compounds antedate silver in photography. Ultraviolet light caused a series of chemical reactions in the solution, leading from the charged ions to a final neutral species $[Pt^{IV}(NH_3)_2Cl_4]$. Bacterial tests of the separated intermediates and final neutral product showed that the latter was the chemical causing filamentation, and was chemically identical to the electrolytically formed agent. A. Thomson, in our laboratory, synthesized the neutral species by known chemical techniques and tested it. It had no activity!

We had only one remaining possibility. The neutral compound exists in two isometric modifications; the trans-$[Pt^{IV}(NH_3)_2Cl_4]$ and the cis-$[Pt^{IV}(NH_3)_2Cl_4]$. The former was the more thermodynamically stable and was the one we first prepared. We now synthesized the cis configuration, and finally, achieved complete success.

Platinum has two dominant valence states, +2 and +4. The lower state forms square planar complexes, and the latter forms octahedral complexes. We now synthesized the +2 complex, and it also was active in forming filaments. Thus the two active chemicals are cis-$[Pt^{II}(NH_3)_2Cl_2]$ and cis-$[Pt^{IV}(NH_3)_2Cl_4]$. These structures are shown in Fig 2. The trans structures have the two similar chemical groups (ligands) on opposite sides of the molecule, and both trans species are inactive at low concentrations (parts per million in solution), but begin to suppress growth at higher concentrations.

Now we had done a strange thing, for by the circuitous route described, we had discovered a compound first synthesized in 1845 and known as Peyrone's Chloride. The molecular structural differences between the cis and trans complexes had been solved by Werner in 1890, who, in so doing, established the basis of modern coordination chemistry. What little value we added by this whole exercise was the use of a biologic test for

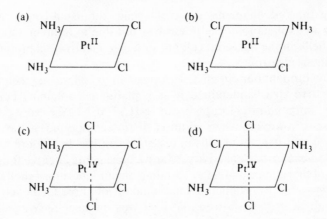

Figure 2 Molecular structures of anticancer active (cis configurations) and nonactive (trans configurations) platinum complexes. (a) *cis*-dichlorodiammineplatinum(II); (b) *trans*-dichlorodiammineplatinum(II); (c) *cis*-tetrachlorodiammineplatinum(IV); (d) *trans*-tetra-chlorodiammineplatinum(IV).

identification of the complex, thus establishing a clear and interesting biologic activity of some coordination complexes of platinum.

2   THE EFFECTS OF PLATINUM COMPLEXES ON BACTERIAL CELLS

Clinical use of metal complexes, particularly of arsenic, antimony and mercury in the treatment of bacterial infections has a long history. The noblest scion was probably Salvarsan, developed by P. Ehrlich about the turn of the century as a specific for syphilis. It was also almost the last of the line. For, in the first half of this century, rapid progress in organic chemistry and biochemistry produced a proliferation of antibacterial drugs, culminating in the enormously successful sulfonamides and finally, the antibiotics. This success fixed the attitude of the next generations of scientists, and metal complexes were largely ignored thereafter.

The antibacterial activity of some platinum group metal complexes was first studied by F. P. Dwyer and his co-workers in 1953. They found the relatively inert chelated complexes of ruthenium with phenanthroline to be quite good bacteriostatic and bacteriocidal agents against Gram-positive microorganisms. Unfortunately, these charged complexes also produced a severe neuromuscular toxicity, 'curare like'' behavior, which limited their use to topical (skin) administration. Limited clinical trials did establish a usefulness for these complexes in the treatment of some skin infections such as dermatosis, dermatomycosis and others, but little further work was done to bring these complexes into general use.

Our laboratory first called attention to the bacterial effects of the simpler complexes in 1965. Over the next few years, in cooperative studies with microbiologists, a number of papers were published describing a multiplicity of effects on microorganisms caused by various complexes of platinum group metals: platinum, palladium, ruthenium, rhodium, osmium, and iridium.

Consider first the filamentation effect. Trials of many complexes established that mainly those complexes which were neutral and had no electrically charged ions in solution, markedly inhibited cell division in bacteria. The cis configuration was active, the trans was not. They did not inhibit growth unless the concentrations were greatly increased. They were associated in the cell primarily with nucleic acids (RNA and DNA) and with some soluble proteins.

Gram-negative rods were the most sensitive to this effect, Gram-positive rods much less so and spherical bacilli (cocci) not at all. Forming a filament was not a terminal event for the bacterium. If the platinum complex was removed from the solution, or the filaments transferred to a normal medium suitable for growth, the filaments begin to divide into normal bacteria, looking much like a string of sausages in the process, growing into quite normal colonies. The division occurs all along the length of the filament and not just at the ends. This was quite a different pattern from the filamentation caused by chemicals such as the nitrogen mustards, where filamentation is a terminal event.

The difference may reside in the fact that the nitrogen mustards block DNA synthesis and each such filament contains only a small number of copies of the genetic information (genome) whereas the platinum complex does not stop new DNA synthesis in bacteria at the concentration causing filaments to appear, and the DNA exists in multiple genome copies as continuous strands or large clumps throughout the filament. This, by the way, is quite different from the effects of these complexes on mammalian cells as discussed below. It is also one of the major differences in the biological effects of the nitrogen mustards, bifunctional alkylating agents and potent anticancer agents, and the platinum complexes.

The complexes which form ions in solution such as $(NH_4)_2[PtCl_4]$, which ionizes to $PtCl_4^{2-}$ and $2(NH_4)^+$ are quite poisonous to the bacteria, causing a large cell kill at low concentrations, and few or no filaments. These ions react with proteins in the cytoplasm of the cell almost exclusively, compared to the strong nucleic acid association of the neutral complexes.

Measurements of these various reactions required a sensitive technique for detecting the minute amounts of platinum incorporated by the cells. This necessitated the use of a radioactive isotope of platinum as a tracer. E. Renshaw and A. Thomson produced the isotope ^{191}Pt by irradiating an iridium foil in the proton beam of the Michigan State University Cyclotron, chemically separating the platinum isotope from the other metals present and synthesizing the charged and the neutral complexes for the bacterial tests. In recent years, the radioactive isotope ^{195m}Pt has been gererously prepared for us at Oak Ridge Laboratories by K. Poggenberg. This also is a γ-ray-emitting isotope with a four day half-life. This means a bout of hectic activity in our laboratory with each delivery to accomplish all our experiments before the level of radioactivity diminishes below our detection threshold.

R. Gillard and his co-workers at Kent University have extended these studies, and shown that organic complexes of rhodium produce similar effects. G. Gale and his associates at the Medical University of South Carolina developed a parallel story of the photochemical transformations and filamentation of bacteria by the cis isomer of the neutral complex with iridium instead of platinum. Thus the experience now accumulated suggests a generality of the bacterial phenomena with the complexes of the other platinum group metals.

Certainly the development of new bacteriocidal agents, particularly those which are active against Gram-negative bacteria, is a very desirable goal. However, the report of anticancer activity of these complexes shifted the weight of research to this more urgent problem. And just as the electric field experiment was bypassed—temporarily I hope—so too the bactericidal utility was put in limbo by more exciting developments. Before moving on to these developments a third bacterial effect needs discussion since it provides some insight into the possible mechanism of how cancers are affected by these complexes.

S. Vasilukova, née Reslova, a young Czechoslovakian microbiologist, and an ex-student of J. Drobnik who contributed much to our microbial experiments, worked with strains of E. coli bacteria that had been previously infected with a bacterial virus (λ-bacteriophage). In these lysogenic bacteria the genetic information of the virus has been incorporated into the cell, but it is repressed so that it is not normally detectable. It replicates during cell division along with the bacterial DNA and so is not lost or diluted out after many divisions.

This is the bacterial equivalent to slow, or latent virus infections in mammals and man. These bacterial strains are called lysogenic, since a number of physical agents such as X-rays or UV light, and some chemicals such as the nitrogen mustards and carcinogens, can derepress the viral genome causing an active viral infection leading to the dissolution—the lysis—of the cell. These effects are easily measured when the bacteria are grown in test tube cultures. The platinum complexes, for example, are added as a few parts per million concentration in the growth medium. The bacteria grow, forming filaments for about three hours, then rather quickly, the milky opacity of the culture diminishes, and in a few hours the culture is water clear; the cells have all lysed. The cis-$[Pt^{II}(NH_3)_2Cl_2]$ complex is extremely efficient in inducing such lysis—less than 0.1 ppm in the culture produces a detectable effect.

It should be pointed out here that all the platinum complexes which are active anticancer agents are also efficient inducers. Those complexes which are not active do not cause lysis. So far, there is a complete isomorphism between the set of active anticancer complexes and the set of efficient inducers. Earlier, we had believed that a good correlation existed between anticancer active complexes and the filament-forming complexes. After a while, however, exceptions in both classes occurred which decreased our faith in this correlation. The correlation of lytic induction and anticancer activity has held up well. In fact, when R. Adamson at the National Cancer Institute reported the anticancer activity of gallium salts, we tested these and, indeed, they did induce lysis in lysogenic bacteria.

We then tested salts of other Group IIIA elements, aluminum and indium, and these too proved to be inducers. It was only after we had predicted, but not published, the activity of aluminum and indium in these tests that Adamson reported them active as anticancer agents. The verification of the prediction tended to reinforce our belief in the correlation, and more important, in the possibility of a similar mechanism of action in the two apparently dissimilar effects.

In 1953, A. Lwoff, had reviewed evidence showing that water soluble mutagens, carcinogens and anticancer drugs were potent inducers of lysogenic bacteria, a strong hint that underlying these four different effects there was a common mechanism, and that it involved an interaction of the causative agent with cellular DNA. The importance of the agent-caused lesion in the DNA in these processes was further enhanced when Vasilukova returned to her native land and performed the experiment called 'indirect induction.' In this case a strain of nonlysogenic bacteria with the sexual transduction factor, F^+ was treated with the platinum complex. These cells were allowed sexually to conjugate with a lysogenic strain, F^-, which had not been treated with the platinum complex. In this process, only a portion of the DNA of the cell is transferred. Yet the recipient cells were induced to lyse. Later on I speculate on the sequence of events arising from this correlation in order to account for the anti-cancer action of the platinum drugs. But first I must carry the story forward to the discovery of their utility as cancer drugs.

3   THE ANTICANCER ACTIVITY OF PLATINUM COMPLEXES

By 1968 we had achieved a certain degree of understanding of the bacteriologic effects of the platinum complexes, and we had synthesized and repeatedly tested the *cis*-dichlorodiammineplatinum(II) which we now took as a model for the active neutral complexes. We were primed to try the chemical against a cancer. The logic was somewhat naïve: the complex stopped cell division in bacteria at concentrations without marked toxicity, perhaps then it would stop cell division in tumors which grow rapidly, without unacceptable toxicity to the host animal. J. Toth-Allen first determined the safe dose levels which could be injected into the peritoneal cavity of mice. The dose which killed 50% of the animals (LD_{50}) was about 13 mg of the drug per kilogram of animal body weight. A dose of 8 mg/kg was nonlethal.

VanCamp then implanted in these mice a standard transplantable animal tumor, the solid Sarcoma-180. This was administered as a 10 mg piece of tumor tissue inoculated beneath the skin under one armpit. The tumor fragment increased its mass about 100 times over the next 10 days. The tumor could be cut out, since it remained localized—nonmetastatic—and weighed. The standard protocols of the National Cancer Institute called for implantation of the tumor on day 0, injection of the drug on day 1, and sacrifice of the animals on day 8. The average tumor size of the treated group is divided by the average tumor size of the untreated group, the negative controls. For a drug to be considered effective against the tumor, the treated to control (T/C) tumor size ratio should be less than 0.5. Our first test values were well below this. We repeated this test more than half a dozen times to be sure it was not a peculiarity due to our inexperience. It was not; each new test reconfirmed the activity.

We also tested a number of other neutral platinum complexes and in this new biological effect, we again saw the stereospecificity that had occurred in the bacterial tests; the cis configurations were active, the corresponding trans configurations were not. The implications of this are significant: it shows that the platinum complexes retain their geometry in the biologic environment, they were not degraded to heavy metal ions which could be nonspecific poisons. The specific chemical reaction leading to the biological effect was sensitive to molecular geometry, and was most likely to involve a macromolecule such as a protein or a nucleic acid. We also were presented with a simple test to determine the significant chemical reaction. Both cis and trans complexes undergo roughly similar, multiple reactions in the cell, but obviously only those reactions which the cis configuration can undergo, but the trans configuration cannot, are likely to be significant.

After confirming these results I contacted G. Zubrod, head of the chemotherapy branch of the National Cancer Institute, and apprized him of the results. I was invited to discuss this with his associates at Bethesda, Maryland. After my short lecture, which was received with perceptible, but understandable coolness, I left samples of the four complexes to be tested in their tumor screen, the L1210 leukemia in mice. A few months later I was informed that the complexes were also active in their system, and it was suggested to me that a grant proposal to pursue this research would not be unfavorably received. A proposal was duly submitted, and approved, and N.C.I. support has continued ever since.

But meanwhile we had tried a variant of the protocols, and this gave us the first hint of the true potency of these complexes. Instead of injecting the complexes on day 1, we waited until the tumor was about a gram in weight (in a 20 g mouse!) and then injected the drugs (on day 8). All the tumors regressed and all the animals were cured. A time sequence photograph of two mice is shown in Fig 3. This was an unusual result since we were not aware of any other anticancer drug capable of regressing large Sarcoma-180 tumors. As would be expected, the surviving animals showed strong immunologic rejections of reimplants of the same tumor up to the longest time tested, 11 months. The animals lived out their full life expectancy; about 30 months, and died of normal, age-related causes.

With confirmed activity against two different animal tumors we were ready to publish the preliminary results, which we did in a short paper in *Nature* in 1969.[1] An American journal of almost equal distinction had turned the manuscript down because a referee had commented that it was not noteworthy since so many new drugs with activity were being found. Indeed, there was a strong possibility that the *Nature* paper would be lost in a crowd of similar reports of new anticancer chemicals that were flooding the literature at the time. It was rescued from potential oblivion by the interest and good graces of Professor Sir Alexander Haddow, then head of the Chester Beatty Institute in London.

Curiously, he had an intuitive feeling that platinum complexes might be effective anticancer agents, and had already tested some earlier, without success. On hearing of our results he had these new complexes synthesized and tested against a different tumor system, a myeloma tumor (ADJ/PC6) in mice, and again confirmed the activity. He wrote to me of

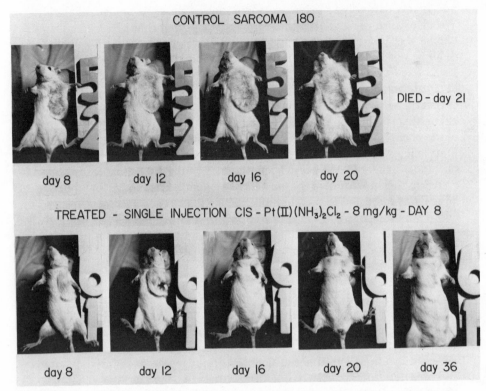

CONTROL SARCOMA 180

DIED - day 21

day 8   day 12   day 16   day 20

TREATED - SINGLE INJECTION CIS - Pt(II)(NH₃)₂Cl₂ - 8 mg/kg - DAY 8

day 8   day 12   day 16   day 20   day 36

Figure 3 Time sequence photographs of two mice with solid Sarcoma 180 tumors. The mouse at the top was an untreated negative control. She died on day 21 when the tumor weighed about 3 g. The bottom mouse was in the group treated on day 8 with an intraperitoneal injection of cis-dichlorodiammineplatinum(II). Her tumor was completely regressed six days after treatment, and she died of age-related causes almost three years later.

the results obtained at the Chester Beatty Institute and extended an invitation to visit with him and some of his colleagues, which I accepted with alacrity. This began a strong cooperative group in Britain which included T. A. Connors and J. J. Roberts of the Chester Beatty Institute, R. J. P. Williams of Oxford University, and M. Tobe at University College, London, all of whom have contributed much to the advance of this new research field.

[Personal comment: We later learned that R. Mason, who had helped in our earlier bacterial studies, had sent some cis-dichlorodiammine-platinum(II) to a friend to test for anticancer activity in 1966. His friend overdosed the animals, they all died, and he reported back that the drug was too toxic! There must surely be a lesson somewhere in this story.]

In recent years cis-dichlorodiammineplatinum(II) has been tested against a wide variety of animal tumors.[2] While this drug is by no means the most active of the platinum complexes, it was the first chosen by the National Cancer Institute to be slated for clinical trials. This fact made studies with this drug more imperative, and most further research making up the bulk of the, by now, over 1000 papers in the field, are concerned with it. A tabulation of the best animal test results is shown in Table 1.

The best results are indicated here solely to convey a qualitative impression of the activity. Most tests were performed with small numbers of animals, making statistical analyses meaningless. Besides, the perversity of animal responses, which indicates a lack of appreciation or knowledge of all the important variables necessary to control, makes either the average or best results of dubious numerical value. Nevertheless, we have compiled a list of 16 tumor types, including transplantable tumors, chemically derived tumors (from carcinogens) and virally derived tumors (from oncogenic viruses). The drug is active against all types. The conclusions that may be drawn from these tests are that the drug:

1. Exhibits marked, rather than marginal antitumor activity.
2. Has a broad spectrum of activity against drug resistant as well as drug sensitive tumors.
3. Is active against slow growing as well as rapidly growing tumors.
4. Is active against tumors normally insensitive to 'S' phase (DNA replicative stage) inhibitors.
5. Regresses transplantable as well as chemically and virally induced tumors.
6. Has shown no animal specificity since it works in mice and rats, either inbred or random bred, and in chickens.
7. Is useful for disseminated (e.g. leukemias) as well as solid (e.g. sarcoma) tumors.
8. Is potent, in that it can rescue animals when injected a few days before death from certain types of tumors.

Thus the credentials of the drug, and by implication, others in the class of platinum group complexes, as an active anticancer agent in animals are well established.

We were then faced with a series of questions, the answers to which were urgently needed, and which required for these answers expert competence in coordination chemistry, biochemistry, biophysics, molecular biology, physiology, pathology, pharmacology, electron microscopy,

Table 1 Best Results of the Antitumor Activity of *cis*-Dichlorodiammineplatinum(II) in Animal Systems

Tumor	Host	Best Results
Sarcoma-180 solid	Swiss white mice	T/C = 2–10%[a]
Sarcoma-180 solid (advanced)	Swiss white mice	100% cures
Sarcoma-180 ascites	Swiss white mice	100% cures
Leukemia L1210	BDF$_1$ mice	% ILS = 379%; 4/10 cures[b]
Primary Lewis lung carcinoma	BDF$_1$ mice	100% inhibition
Ehrlich ascites	BALB/c mice	% ILS = 300%
Walker 256 carcino-sarcoma (advanced)	Fisher 344 rats	100% cures; T.I > 50[c]
Dunning leukemia (advanced)	Fisher 344 rats	100% cures
P388 lymphocytic leukemia	BDF$_1$ mice	% ILS = 533%[b]; 6/10 cures
Reticulum cell sarcoma	C+ mice	% ILS = 141%[b]
B–16 melano-carcinoma	BDF$_1$ mice	% ILS = 279%[b]; 8/10 cures
ADJ/PC6	BALB/c mice	100% cures; T.I. = 8[c]
AK leukemia (lymphoma)	AKR/LW mice	% ILS = 225%[b]; 3/10 cures
Ependymoblastoma	C57BL/6 mice	% ILS = 141%[b]; 1/6 cures
Rous sarcoma (advanced)	15-I chickens	65% cures
DMBA-induced mammary carcinoma	Sprague Dawley rats	77% total regressions
		3/9 free of all tumors
ICI 42, 464-induced myeloid and lymphatic leukemias	Alderly Park rats	% ILS = 400%[b]

[a] T/C = $\dfrac{\text{tumor mass in treated animals}}{\text{tumor mass in control animals}} \times 100$.

[b] % ILS = % increase in lifespan of treated over control animals.

[c] TI = therapeutic index (LD_{50}/ED_{90}), ED_{90} = effective dose to inhibit tumors by 90%.

immunology, and finally, clinical medicine. In short, the entire panoply of disciplines in chemistry and biology were needed. We alone could not do it, nor could any small group of laboratories. A worldwide network of cooperating laboratories was called for, and established. They were supported by public funds, cancer societies, and to a generous degree, the platinum industry.

[Personal comment: I recall two strict admonitions from my major professor when I informed him of my growing interest in biophysics. These were, not to work with medical doctors untrained in research and to avoid cancer research, since many had tarnished their reputations from a malignant neglect of scientific objectivity in their desire to do something useful. I have broken both injunctions, but I cannot say that I am sorry. With a very few exceptions, all connected with the network impressed me as dedicated, selfless, humane scientists. The expected ego clashes and political infighting that characterize so much of science seems to have been muted by the urgency of the problem at hand.]

4 MOLECULAR STRUCTURE DETERMINES THE ANTICANCER ACTIVITY

Of the myriad questions arising from this discovery of the anticancer activity in mice of some platinum complexes, one which we did feel competent to attack, particularly since we had the advice of some very able inorganic chemists, R. Mason of Sussex University, R. J. P. Williams of Oxford University and M. Tobe of University College, London, was the so called structure—activity relation. Without requiring detailed knowledge of the molecular interactions of the chemicals in the biologic system, we simply induced structural changes in the molecules by known synthetic techniques and tested them against a standard mouse tumor. If many variations are tried, then a catalogue of these chemicals, with a simple numerical measure of their activity, should exhibit some regularities. This allows them to be grouped in subclasses, and each subclass can be analyzed for common chemical properties. The more chemicals that are tested, the sharper will be the subclassification. However, a reasonable limit had to be set for these syntheses.

As an example, one starts with one metal of the platinum groups, platinum itself. It has the two major valence states $+2$ and $+4$. Take the latter. It can associate in an octahedral complex with six ligands, atomic groups bound to the metal. The individual ligands may be chosen from a large group, but let us restrict it to just 10. Therefore, for this one metal valence state we have about one million potential variations. Obviously, the required manpower to synthesize, purify and characterize such numbers of chemicals is beyond the world's capacity even if all laboratories were recruited for this sole purpose, to say nothing of the 30 million mice required.

A drastic compromise was called for, and here the intuition of the experts in coordination chemistry was essential. Actually, only about 1000 complexes have been studied in various laboratories. Of these, about 10 to 20% are active. The molecular structures of some of the most active complexes are shown in Fig 4. The number of 'actives' is this large simply because most have started with a known active complex and produced small variations upon it. In a large morass of chemicals, we have found, through luck or cleverness, a few small islands of success and we stray far from these only at some peril. It is not a very satisfying situation when endangered grant renewal is the penalty for boldness.

Nevertheless, in the areas explored, some common features have emerged which link structure to activity. We embody these here in a set of 'rules of thumb,' since they can hardly lay claim to general validity.[3] These are:

1. The complexes exchange only some of their ligands quickly in reactions with biological molecules.

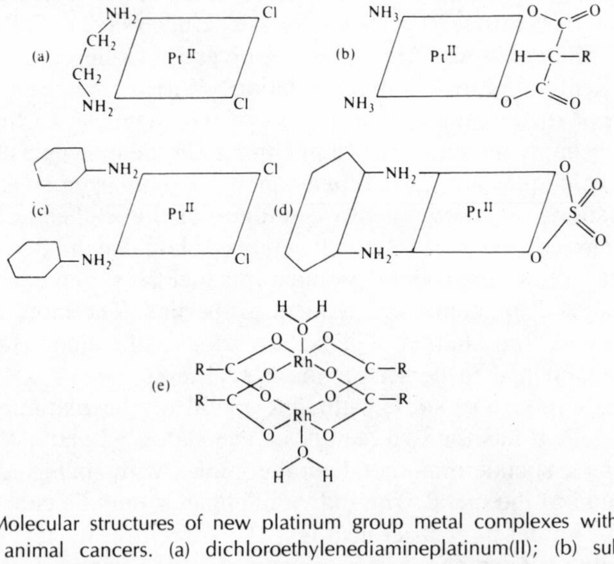

Figure 4 Molecular structures of new platinum group metal complexes with high activity against animal cancers. (a) dichloroethylenediamineplatinum(II); (b) substituted (R) malonatodiammineplatinum(II); (c) cis-dichlorobis(cyclohexylamine)platinum(II); (d) sulfato-1,2-diaminocyclohexaneplatinum(II); (e) rhodium(II) carboxylate.

2. The complexes should be electrically neutral, although the active form may be charged after undergoing ligand exchanges in the animal.

3. The geometry of the complexes are either square planar or octahedral.

4. Two cis monodentate or one bidentate leaving group (exchangeable ligands) are required; the corresponding trans isomers of the monodentate leaving groups are generally inactive.

5. The rates of exchange of these groups should fall into a restricted region, since too high a reactivity will mean that the chemical reacts immediately with blood constituents and never get to the tumor cells, while too low a reactivity would allow it to get to the cells, but they would do nothing once there.

6. The leaving groups should be approximately 3.4 Å apart on the molecule (an interesting number, since the spacing between the steps of the Watson—Crick DNA ladder is also 3.4 Å).

7. The groups across the molecule from the leaving groups should be strongly bonded, relatively inert amine type systems.

We certainly do not intend these rules to restrict future research, but only to encompass a large amount of past experience with platinum(II) complexes. Obviously exceptions will, and have already, occurred. For example, the high activity of bidentate leaving groups such as oxalate and malonate (see structures of Fig. 4) first synthesized by M. Cleare and J. Hoeschele in this laboratory are not encompassed; nor is the effect of cyclic amines, developed by Tobe, which decrease the solubility of the complexes, but markedly enhance the antitumor activity. Here, studies of the relative solubilities in oil and water, the partition coefficient, may be significant in determining activity.

We can now present a broad outline of the fate of the drug, *cis*-dichlorodiammineplatinum(II), after injection into the peritoneal cavity of the mouse. Within minutes the drug leaves the cavity through the blood and lymph circulation. The high chloride concentration of these extracellular fluids prevents the chlorides from leaving the molecule, thus maintaining the structural integrity. The intact drug is rapidly excreted in the urine, with a half-life in the body of about one hour. The excreted drug is 95% the unchanged molecule but about 5% is attached to proteins. The drug is passively transported across the cellular membrane—no active transport (carrier) is necessary. Once inside the cell, the lower chloride content of the cytoplasm ($\frac{1}{30}$ of that outside the cell) allows the chloride to exchange with water according to the following scheme:

$$[Pt^{II}(NH_3)_2Cl_2]^\circ + H_2O \rightleftharpoons [Pt^{II}(NH_3)_2(H_2O)Cl]^+ + Cl^-$$

$$[Pt^{II}(NH_3)_2(H_2O)Cl]^+ + H_2O \rightleftharpoons [Pt^{II}(NH_2)_2(H_2O)_2]^{2+} + Cl^-$$

Depending on the hydrogen ion concentration, the H_2O may be changed to $(OH)^-$. This aquated species reacts primarily with the nitrogens of the DNA bases leading to the primary lesion responsible for the anticancer effect. While the formula for the diaquo species implies a simple, single structure, we have recently discovered that it is slightly more complicated than that. In fact, isolation of crystal species of the diaquo complex under slightly different conditions have yielded one monomer, one hydroxy bridged dimer, two hydroxy bridged trimers, a tetramer and two other not yet resolved crystal forms. This emerged from a cooperative study between B. Lippert of this laboratory and C. J. L. Lock of McMaster University. It is not yet clear what role, if any, these various structures have in the anticancer activity or toxicity of the parent drug.

I have been purposely nebulous so far on where the DNA is, and what sites of the DNA are involved. In order to be more concrete we require a short description of the molecular biology studies.

5   WHAT DO THE PLATINUM COMPLEXES DO TO MAMMALIAN CELLS?

Two options are available generally to study the effects of drugs on cells; first, inject the drug in animals, excise the desired cells and examine these for changes, the in vivo system; second, use purified cells growing in tissue culture, the in vitro system. The former is more relevant, but the latter is scientifically 'cleaner.' Both should be done, and in the case of the platinum drugs, were simultaneously and independently performed in two laboratories. Both the techniques were used by Gale and his associates at the Medical University of South Carolina, while only the second was done in our laboratory. There is general agreement on the results obtained by the two methods.[4]

By the use of radioactively labeled precursor chemicals, the cell's ability to synthesize macromolecules such as DNA, RNA, and proteins, after treatment with the drug, can be measured. A typical result is shown in Fig 5, for exposure of the cells to the equivalent level of drug found in tumor tissue of a treated animal. The synthesis of new total DNA is selectively and persistently inhibited. Total RNA and protein syntheses are not markedly affected until much higher drug dose levels, which are frankly toxic to the cells, are used. The level of inhibition of DNA synthesis is dose dependent. Its onset is slow, taking about 4 to 6 h after drug exposure to reach a nadir. It was surprising that there was not a large cell kill at the therapeutic dose level. The cells first grew into giant cells which, after a few days, showed the appearance of many nuclei and

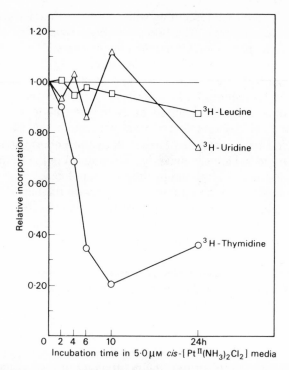

Incubation time in $5{\cdot}0\,\mu\text{M}$ cis-$[\text{Pt}^{\text{II}}(\text{NH}_3)_2\text{Cl}_2]$ media

Figure 5 Effects of cis-dichlorodiammineplatinum(II) on macromolecular syntheses in human amniotic cells in tissue culture at a concentration ($5\,\mu\text{M} \simeq 1$ ppm) similar to that found in tumor cells in animals treated with a therapeutic dose of the drug. DNA synthesis is measured by the incorporation of radioactively labeled thymidine and is severely and persistently inhibited. The synthesis of RNA, measured by radioactive uridine, and protein, measured by radioactive leucine, is not significantly different from control (nontreated) cells represented by the horizontal bar at 1.00.

eventually divided into a number of single cells. I will return to this important result later.

H. Harder, then in our laboratory but now at Oral Roberts University, was responsible for these studies. He also checked that the synthesis of precursor molecules for the DNA, and the transport of these across membranes was not responsible for the inhibition. He has more recently shown that the ability of the DNA to act as a template for new synthesis is strongly inhibited by the platinum drug. These results can be most reasonably explained by the hypothesis that the anticancer activity of the platinum drugs arises from a primary attack on DNA. The battle to discover the molecular mechanism of action was therefore joined on the field of metal complex interactions with DNA, and numerous other laboratories entered the fray. The booty has been rich, embarrasingly so.

We now know that *cis*-dichlorodiammineplatinum(II) can crosslink the two strands of the double helix of DNA, an exciting discovery since this type of linkage had already been invoked to account for the anticancer activity of the bifunctional alkylating agents such as the nitrogen mustards. It was made almost simultaneously in three laboratories, but most elegantly by J. J. Roberts and his co-workers at the Chester Beatty Institute. It can also, apparently, crosslink two neighboring bases stacked on a single strand, which significantly, the complex in the trans configuration should not do. The platinum drug does not react with the sugar–phosphate backbone of the strands, but only with the bases. Nor does it appear to intercalate between the bases. It reacts most strongly with the G—C rich regions of DNA and can, through the technique of gradient centrifugation, be used to characterize the relative G—C/A—T content of DNA.

The platinum complex—DNA reaction is very slowly reversible in vitro, but it may be removed more rapidly within the cell by the actions of DNA repair enzymes. The two available exchangeable groups of the platinum can react at two sites on a given base (primarily the purines, guanine and adenine), or with single sites on two different bases, or finally a single ligand of platinum exchanges at only one site of a base. It will take a considerable period to sort out the multiplicity of such reactions and to identify finally one or more as the necessary lesion for anticancer activity. In the meantime, it is clear to many of us that metal complex interactions with nucleic acids are too poorly understood, and too important to remain so.

It must surely be nagging the reader by now, as it has us for some years, that in all of the above work no clear distinction has emerged between effects on tumor cells and on normal cells. The same effects qualitatively, and very likely, quantitatively appear in many cell types, and yet we have suggested that these effects are the primary lesion leading to anticancer activity. Chemical studies of DNA are no doubt important, but since we cannot say in what way DNA differs in cancer cells from normal cells, we cannot answer the question of why the cancers are killed and not the animals. The question is by no means trivial. It cuts to the heart of cancer chemotherapy.

Justifiably, some people are unhappy that we have not yet discovered drugs with higher curative power against cancer. However, many researchers, myself included, are surprised that we have discovered so many that are so good, because we do not know why. All cancer drugs are cellular poisons, but not all cellular poisons are cancer drugs. If the drugs were not eventually more poisonous to cancer cells than normal cells, we would not be injecting them into patients. With the possible exception of L-asparaginase, we have not yet been able to seize upon a unique, exploitable characteristic of cancer cells to produce specific, or even selective tumor cell kill. Yet in the host animal this can, and does, occur with presently useable drugs. This, admittedly simplistic, logic leads naturally to the invocation of the host–tumor interaction rather than the drug–tumor interaction as the source of specificity in the anticancer activity of drugs. Such specificity is usually associated with the host's immune response.

6 HOW DOES SELECTIVE CANCER DESTRUCTION OCCUR?

Here I should like to touch on a more speculative side of the research. We had to face up to the strong evidence at the molecular level that the platinum drug produced a lesion on the DNA of cells, which did not necessarily lead to cell death, and in any case, was not restricted to cancer cells alone, and the final clinical observation that the cancers disappeared in the animal, without unacceptable side effects. There is a wide gap between molecular biology and clinical results. Could we bridge some or all of it with testable hypotheses?

The specificity of the cure is a good clue, since as we noted, specificity is usually associated with the host's immunologic responses. Is there any evidence for such responses of the host? There is, but this evidence consists mainly of an accumulation of weak arguments which cannot be summed to make a strong argument. Briefly, these are: our earliest screening studies of coordination complexes brought to light the peculiar result that some complexes increased the rate of tumor growth by about 200% compared to untreated controls. This is consistent with the already established suggestions that the host animal exerts some constraint on the growing tumor through immunologic reactions, and if these constraints are inhibited (by immunosuppressive agents) without these agents simultaneously exerting antitumor activity, then increased tumor growth rates are expected.

The second involvement of the immune system occurred when we were able to cure large solid Sarcoma-180 tumors in ICR mice. The cured animals rejected any new attempt to reimplant this tumor up to 11 months later. They have obviously developed a heightened immunologic reactivity for this tumor. Interestingly enough, the cure of small tumors did not produce such an immunologic rejection reaction.

This experiment also produced a third unexpected result. It has been accepted, since the classic work of H. E. Skipper and co-workers at the Southern Research Institute, that at least for leukemia, a given dose of a drug kills a constant fraction of the tumor cells present, in fact a first order kinetic process. Yet we are able to cure small tumors and large tumors with a given optimal dose of the platinum drug but not intermediate sized tumors. This is not sensible if one considers direct cell kill only. Similarly, if the optimal dose cures the large tumors, then a much smaller dose should cure the smaller tumors. This, too, is contrary to our experiments. Something other than direct chemical cell kill must be operating to achieve cures.

Dead Sarcoma-180 cells injected into mice do not cause tumors, neither do they induce an immune reaction to reimplanted live tumor cells. Here one must be cautious since only small numbers of live implanted cells (~40) can eventually lead to large tumors and death. But cells treated with the platinum drug at low concentrations, 100 times less than the concentration required to produce extensive cell kill, implanted in the mice do not produce tumors, but do induce an immunologic rejection of pristine tumor cells implanted two weeks later. This experiment is difficult to interpret without invoking the immune system of the host in causing tumor cell death.

If the immune system is involved, then one could anticipate that modulating the host's immune competence should modulate the anticancer activity of the drug. Preliminary experiments by P. Conran in this laboratory, and now at the University of Connecticut, suggest that this is true. Decreasing the immunocompetence of mice by hydrocortisone injections decreases the cure rate of the platinum drug against Sarcoma-180 in ICR mice; while conversely, the nonspecific immune stimulant, zymosan, increases the cure rate against the Sarcoma-180 in BALB/c mice. Unfortunately these systems are not as 'immunologically clean' as one would like, so the experiments are now being repeated using acceptable systems both in our, and Conran's, laboratories.

One of Tobe's complexes, cis-dichlorobis(cyclohexylamine)platinum(II) was tested by T. A. Connors against the ADJ/PC6 myeloma tumor in mice. It cured the tumors completely at a dose 1/500 of the LD_{50}. Such specificity, especially in the absence of any evidence of selective tumor uptake of the drug, is utterly inconsistent with the direct cell toxicity hypothesis. A host response must again be invoked, one with high specificity. And again, the immune system alone has that characteristic.

Finally, we can count the number of tumor cells as a function of time in the animal after a platinum drug treatment known to produce large percent cures. The Ascites Sarcoma-180 in ICR mice is ideal for this purpose. We inject four million cells into the peritoneal cavity on day 0. They

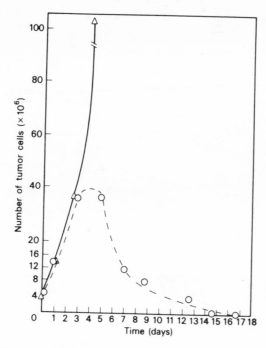

Figure 6 Growth of ascites Sarcoma 180 tumor cells in untreated ICR mice (solid line) and in treated mice (dashed line). Treatment consisted of five injections of 1.5 mg/kg given on day 1 after inoculation of 4,000,000 tumor cells on day 0. The number of cells continues to increase by repeated cell division up to day 4 and then slowly decreases to zero cells (cured animals).

multiply rapidly to two thousand million cells 15 days later, remaining localized in the cavity and killing 100% of the animals. Now we inject the platinum drug on day 1, and sacrifice small numbers of the animals every day. We wash, clean, and count the tumor cells. The results are shown in Fig 6. The cells divide about 2 to 4 times, increasing in number up to 40 million cells by days 4 to 5, before a turnaround occurs and the cell number drops to 0 on days 9 to 10. If direct tumor cell kill by the drug was operative, we should expect a fast decline from about eight million cells to 0 on day 2, with no further cell divisions and no continued growth. This is contrary to the experimental results and again suggests a host mechanism for tumor cell destruction.

We propose that these arguments are consistent with, but do not corroborate, the hypothesis that the platinum drug enhances the antigenic character of the tumor cells, tipping the balance in favor of the host's immunologic intervention to destroy the cancer. The question then be-

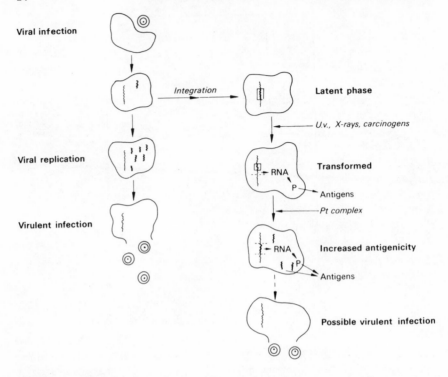

Figure 7 Schematic diagram of one possible hypothesis of the molecular action of *cis*-dichlorodiammineplatinum(II) leading to an enhanced antigenicity of tumor cells by the derepression of virally coded information latent in the cell.

comes, 'how does the platinum drug accomplish this?' Now we return to the derepression story based on Reslova's work. In brief, Fig 7 outlines one potential sequence of molecular events in a mammalian cell which could produce the desired result.

Without making a definite commitment, let us assume the hypothesis that expression of viral DNA is the causative factor in the cell transformation to a cancer state. There is certainly a significant body of experiments indicating that this is true in many mammals, but solid evidence in humans still eludes us. The viral genome, incorporated in the cellular genome, is completely repressed for long periods compared to most cells division times. As in the case of lysogenic bacteria, a wide variety of chemical and physical agents are able to cause a depression of a small part of this latent viral genome, enough say to code the production of one or two proteins. These proteins transform the cell. The existence of temperature sensitive mutants for cell transformation tell us that the production of as little as one virally coded protein is a necessary (but not sufficient) cause of cell transformation. This small number of proteins is also the cause of the antigenicity of the tumor cell. The chemical and physical agents causing derepression are, therefore, carcinogens. If now we add the platinum drug to the cell, it effectively derepresses a larger fraction or all of the viral genomic information. This inevitably leads to the production of a larger number and variety of proteins, and this enhances the antigenicity of the cell.

This scheme, incidentally, provides a simple explanation of 'Haddow's paradox,' that is, certain classes of chemicals and agents that cause cancer, can also cure cancer. The difference between cause and cure indicated here is simply a quantitative one. It is the amount of derepression of the viral genome.

Some experimental information consistent with our hypothesis exists. V. Vonka and co-workers at the Institute of Sera and Vaccines in Prague were able to induce an up to 300% increase in the number of Epstein Barr virus (an oncogenic Herpes type virus) positive cells in a culture of Burkitt lymphoma cells (EB3) by treatment with *cis*-dichlorodiammineplatinum(II). The induction of the new, virus associated, antigens was monitored both by an indirect immunofluorescence test for the coat proteins of the virus appearing at the cell surface, and by the visualization of virus like particles in the treated cells by electron microscopy.

Thus the platinum drug, at least in this case, causes the hypothesized derepression in a cancer cell line, and has enhanced the antigenicity of the cells. While the enhanced antigenicity hypothesis is consistent with a large body of information, and does bridge the gap between molecular events involving the platinum drug interaction with cellular DNA and the host immunologic intervention, it still leaves unexplained the detailed mechanisms of derepression of latent viral genomes, its role in cell transformation and the nature of the immune response. Fortunately, many laboratories are working on these particular problems.

7   HOW EFFECTIVE ARE THE PLATINUM DRUGS AGAINST HUMAN CANCER?

The end of the story must be the answer to this question, for if the drugs are no improvement over present drug therapies, coordination complexes as anticancer agents will pass into history and be forgotten. Happily the emerging clinical results do suggest a significant treatment improvement in, at least, a number of types of cancers.[5]

The first platinum drugs entered human clinical trials in 1972 after a series of preclinical studies of the toxic side effects of different dose levels in dogs and monkeys. These studies established a safe low starting dose, and predicted certain toxicities which would occur when the dose level was escalated. The most severe side effect was the damage to the kidney, with less stringent toxicity to the blood forming elements of the bone marrow, intense nausea and vomiting (which we have good reason to believe is due to a central nervous system effect and not to gastrointestinal damage), and some destruction of the sound detecting hair cells of the organ of Corti in the inner ear leading to transient high frequency hearing loss, and in a few cases, total deafness. The kidney damage is a severe limiting side effect on the allowed dose levels. Nevertheless, below these dose levels anticancer activity was reported.

The patients entered into first clinical trials of a new drug are usually terminally ill with cancers that are no longer responsive to any treatment, and who have signed an informed consent agreement. It cannot be expected that such patients, whose physiologic vitality has been severely sapped both by the illness and the subsequent therapy, particularly with drugs which are immunosuppressive, will show marked responses to the new drug. But invaluable information on the side effects in humans, and appropriate dose levels and schedules can be obtained. In the case of the platinum drug, the predicted side effects occurred plus the hearing loss effect, which could not be easily predicted from animal studies, just as the nausea and vomiting effect which was first found in dogs and monkeys could not have been predicted from all our mice and rat studies, since these latter animals cannot vomit.

Still some patients did benefit from the treatment, with partial or complete remissions in about 20% of the patients, for various time durations. It was noted early on that the drug seemed to produce more responses in patients with genitourinary cancers, ~80% for testicular cancers, >90% for ovarian carcinomas, head and neck cancers (~40%), and some lymphomas (~40%), than in other types of cancer. These preliminary results tend to augment the impression that the platinum drug is active against a fairly broad spectrum of cancers. One total failure noted was colon carcinoma, a very drug resistant cancer. Even here, however, a bright note occurred when the platinum drug is used in combination chemotherapy as described later.

At this earlier stage, a fair estimate of the clinician's attitude toward the platinum drug was encapsulated in the remark, ". . . [it] appears to be too good a therapeutic agent to abandon, yet too toxic for general use." The danger referred to was the kidney toxicity, an unfamiliar side effect with the older cancer drugs.

In the second phase of the clinical trials, the drug was used on early patients in a crossover pattern with other drugs. Those patients not responding to the first drug were subsequently treated with the second drug and vice versa. I should point out that although various phases and protocols of testing are neatly blocked out by the National Cancer Institute in the United States, these delineations tend to be somewhat blurred in actual practice since the medical doctor's first priority is to benefit the patient rather than exactly fulfill an experimental protocol. Previous experience with other drugs in Phase II tests generally produced higher response rates, and the platinum drug was not an exception.

Chemotherapy of cancer with single agents has not yet proven to be of significant long-term value. Within this last decade large improvements in response rates occurred when drugs were administered in combination. The desiderata for such combinations are synergism, or at least additivity, of anticancer effects, and less than additivity for any one type of dose limiting side effect. A number of such three-, four- and even five-drug combinations has been developed with encouraging results in patients, but true synergism is relatively rare. The platinum drug was reported by J. Venditti and his co-workers at the National Cancer Institute to produce synergistic action in animal tumor systems with a large number of other anti-cancer drugs. This was verified shortly thereafter by R. Speer and H. Ridgeway at the Wadley Institute of Molecular Medicine in Dallas, Texas, and a number of other laboratories including our own. This provided the necessary rationale for combination therapy with the platinum drugs.

The first success was found by F. Ansfield and R. A. Ellerby at the University of Wisconsin. They treated terminally ill colon carcinoma patients with the *cis*-dichlorodiammineplatinum(II) drug and 5-fluorouracil. Three of nine such patients showed objective (greater than 50% reduction of tumor masses) responses for varying durations. This is presently under further study at the Mayo Clinic in Rochester, Minnesota.

The studies of J. Wallace, J. Holland, and co-workers at the Roswell Park Memorial Institute in Buffalo, New York, on testicular cancers probably saved the platinum drug from premature burial. They reported responses in 13 of 15 patients, seven of which were complete, and of long duration. Here again another frustration in cancer research was manifest. Drugs do exist which can, in a large percentage of cases, cause complete regressions——disappearance of all detectable symptoms of cancers, but only for limited times. The tumor then returns with a vengeance and is usually refractory to further treatments. Why not cures?

A cure is defined by the fact that the patient achieves a normal life expectancy and dies of causes not related to the cancer. Operationally, however, this cannot be an acceptable definition. Instead, experience dictates that if a given cancer does not return within some specified time (usually about 5 years) the patient is considered cured. Therefore, improved therapies which push the survival rates further out in time could cross the threshold to produce cures. This may, in the near future, occur with testicular cancers.

E. Cvitkovic, I. Krakoff, and co-workers at the Sloan–Kettering Memorial Center for Cancer and Allied Diseases in New York, made the next push toward this goal. Cvitkovic reasoned that heavy metal kidney toxicity can be ameliorated by administering an osmotic diuretic agent such as D-mannitol, while hydrating the patient. Since platinum is a heavy metal causing kidney toxicity, why not try this treatment? While the logic is not totally defensible, his experimental results were effective—he was able to increase the safely administered dose of the platinum drug by 300%, with a concommitant increase in efficacy. The platinum drug was then added to their standard multiple drug therapy which, on average, put about 50% of the patients in complete remission for a median life expectancy of six months. The new combination has now produced greater than 90% complete remissions, whose mean duration cannot be established since in the succeeding 14 months there have been little or no relapses. Similar good responses in the treatment of testicular cancers with the platinum drug in combination therapy were also reported by L. Einhorn and co-workers at Indiana University Medical Center in Indianapolis, Indiana, and by C. Merrin at Roswell Park Memorial Institute.

This optimistic progress towards achieving complete remissions in large percentages of the patients and for increasing duration is now happening in treatment of ovarian carcinoma. The first indication of activity of the platinum drug in this cancer was reported by E. Wiltshaw of the Royal Marsden Hospital in London in 1973. Again, combination therapy has proved more efficacious than single agents. H. Bruckner, J. Holland and associates at Mount Sinai School of Medicine in New York developed a combination of the platinum drug with adriamycin, and more recently have added cyclophosphamide, and are reporting response rates of 70% and better, with longer durations of response.

Cancers of the head and neck are now treated quite effectively by a combination chemotherapy utilizing the high-dose platinum drug, with bleomycin infusion, as developed at the Memorial Sloan–Kettering Cancer Center by R. E. Wittes and his co-workers. More recently, A. Yagoda and associates at the same institution have shown the marked efficacy against bladder carcinomas of the platinum drug, alone and in combination chemotherapy. And finally, both Merrin's group and Holland's group have now added prostrate carcinoma to this growing list of responsive cancers.

It is now certain that combination therapies with the platinum drug are a safe, effective treatment for a number of different types of cancer. This will undoubtedly be extended in the future to a number of other cancers; to a variation of therapies to improve the present results; and finally, to include second- and third-generation platinum coordination complexes which are now known, from animal tests, to be much superior to *cis*-dichlorodiammineplatinum(II).

REFERENCES

1. B. Rosenberg, L. VanCamp, J. E. Trosko, and V. H. Mansour, *Nature (London)* **222**, 385 (1969).
2. B. Rosenberg, *Naturwissenschaften* **60**, 399 (1973).
3. A. J. Thomson, R. J. P. Williams, and S. Reslova, *Struct. Bonding (Berlin)* **11**, 1 (1972).
4. T. A. Connors and J. J. Roberts, Eds., *Platinum Coordination Complexes in Cancer Chemotherapy,* Srpinger-Verlag, New York, 1974, pp. 79–97.
5. Recent clinical results are described in The Proceedings of the III International Symposium on Platinum Coordination Complexes in Cancer Chemotherapy, published in *J. Clin. Hematol. Oncol.* **7,** Parts 1 and 2 (January 1977).

CHAPTER **2**

Heavy Metal
Interactions
with Nucleic Acids

JACQUELINE K. BARTON and STEPHEN J. LIPPARD

Department of Chemistry, Columbia University
New York, New York

CONTENTS

1 INTRODUCTION

The study of heavy metal binding to nucleic acids and their constituents is a burgeoning field of research. Because of their high electron densities, heavy metals are commonly used to stain biopolymers for electron-microscopic examination. A current goal is to develop the chemistry of heavy metal reagents as stains for specific sites on polynucleotides. Electron-dense metals are also used as isomorphous replacement reagents to help determine the phases in X-ray crystallographic studies of macromolecules. The high density of heavy metal reagents also facilitates the separation and isolation of labeled polynucleotides based on their altered sedimentation properties. Fluorescent lanthanide cations afford spectroscopic probes of polynucleotide structure, with energy transfer between bound lanthanides having the potential to reveal structural information about the macromolecule in solution.

The discovery of the antitumor platinum drugs (1) has further stimulated research into the chemistry of heavy metals with nucleic acids and their constituents. While the cellular target of *cis*-dichlorodiammineplatinum(II), *cis*-DDP, is as yet uncertain, a primary response to the drug is the cessation of DNA synthesis. The complex may act by binding to the polynucleotide, disrupting its template activity for the synthesis of new DNA and thereby inhibiting cell division. An understanding of the specific mode and site of action of platinum antineoplastic agents could provide for the design and synthesis of even more effective antitumor drugs.

The design of selective labels for polynucleotides and the description of the interaction of *cis*-DDP with its biological target have in common the need for a fundamental understanding of the coordination chemistry of heavy metals with nucleic acids. Why, for example, does *cis*-DDP exhibit a high index of antitumor activity while the analogous trans isomer is ineffective? What reagents and experimental protocols are required for the stable attachment of heavy metals to specific sites on the polymer without disruptive side reactions or the use of conditions that degrade the polynucleotide? A number of laboratories have approached these problems by studying the reactions of heavy metals with nucleic acid constituents in an attempt to model the more complex interactions that occur with the polymers. Spectroscopic and structural studies have revealed the primary nucleotide binding sites and the factors that influence metal reactivity. The intriguing complexes that have been thus far isolated establish, if nothing more, the diversified role of nucleotides as ligands.

In this chapter our aim is to provide an anecdotal description of current research into the binding of heavy metals to nucleic acids. It is not our intent to provide an exhaustive review of the literature. A number of excellent articles have recently appeared covering selected aspects of this field (2–8). We begin with a discussion of the metal binding sites on monomeric nucleic acid constituents and the complex products that form. Next we examine the binding modes of heavy metals to polymers. The polynucleotide structure imposes conformational constraints that largely determine the stereochemistry of the interaction. Cases where the heavy metal reagent serves as a useful probe of biological structure are examined. The use of paramagnetic transition metal ions in biochemical assays is not stressed, our emphasis being centered primarily on those experiments that utilize the high electron density of heavy metal reagents. Topics covered in detail in other chapters of this book are described only briefly. Finally, proposals for the mode of action of *cis*-DDP are considered. Evidence from both in vivo and in vitro experiments as well as proposed models for the specific geometry at the interaction site are described.

2 METAL BINDING TO NUCLEOTIDES, NUCLEOSIDES, AND THEIR CONSTITUENT BASES

2.1 Properties of Mononucleotides

2.1.1 Nomenclature and Conformations. The mononucleotides, building blocks of DNA and RNA, offer a rich variety of sites for metal reactivity. Each mononucleotide is composed of a negatively charged phosphate unit, the site of electrostatic interactions; a nucleic acid base, the most common site of metal coordination; and a ribose sugar ring. In a polymer it is the ribose group that holds the bases to the chain, determining their relative orientations and therefore the accessibility to metal attack.

The common constituent bases and their numbering schemes are shown in Figure 1. Uracil, thymine, and cytosine are composed of six-membered, heterocyclic pyrimidine rings that differ in their exocyclic substituents. Uracil, found in RNA, and thymine, found in DNA, differ only in the presence or absence of a methyl group at C-5. The purines, guanine and adenine, are composed of two fused heterocyclic rings, a six-membered pyrimidine and a five-membered imidazole. The numbering scheme is clockwise for pyrimidines and counterclockwise for purines when drawn as shown. Less common bases such as 4-thiouracil (s^4U) and hypoxanthine have also been studied as ligands. Thiolated derivatives of the bases such as s^4U form stable bonds with soft metal atoms. Hypoxanthine provides a simpler analog of guanine, lacking the 2-amino substituent.

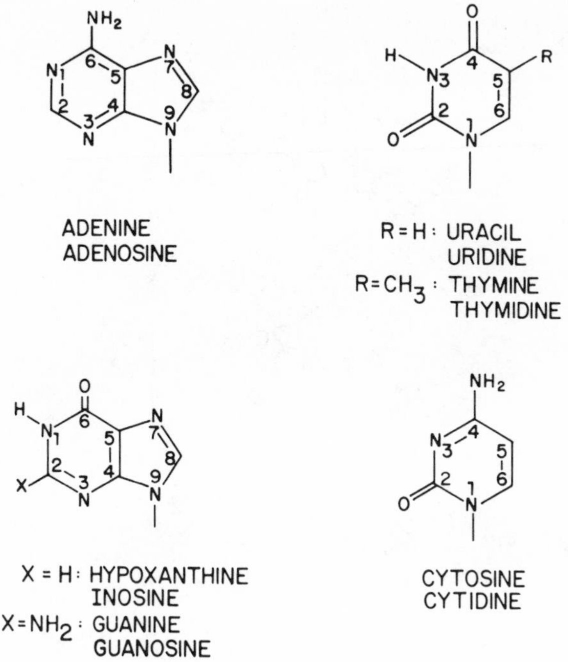

ADENINE
ADENOSINE

R = H : URACIL
        URIDINE
R = CH$_3$: THYMINE
        THYMIDINE

X = H : HYPOXANTHINE
        INOSINE
X = NH$_2$: GUANINE
        GUANOSINE

CYTOSINE
CYTIDINE

Figure 1 Drawings and atom labeling schemes for the common nucleoside bases. The bonding sites to the sugar rings are designated by vertical lines. The upper name is that of the free base. The lower name is that of the ribonucleoside.

Ribose derivatives of the bases that occur naturally in RNA are referred to as ribonucleosides. They include adenosine, guanosine, cytidine, and uridine. The hypoxanthine nucleoside is called inosine. The analogous DNA components are 2'-deoxyribonucleosides, commonly deoxyadenosine, deoxyguanosine, deoxycytidine, and thymidine. Primed numbers are used for the carbon atoms of the sugar and their substituents. The sugars of RNA and DNA differ only in the presence or absence of the 2'-hydroxyl group. Attachment of the base to the sugar C-1' carbon atom is through N-1 for pyrimidines and N-9 for purines. Only pseudouridine, found in tRNA, is bound to the ribose through the C-5 carbon rather than N-1.

Nucleotides are the phosphate esters of nucleosides, the phosphate unit being attached either through the 2'-, 3'-, or 5'-hydroxyl oxygen atom. Mononucleotides linked to one another through 5'- and 3'-sugar–phosphate bonds form the polynucleotide chain. The positive direction of progress is conventionally taken from 5'- to 3'-terminals. In referring to nucleic acids, some common abbreviations will be used: AMP, ADP, and

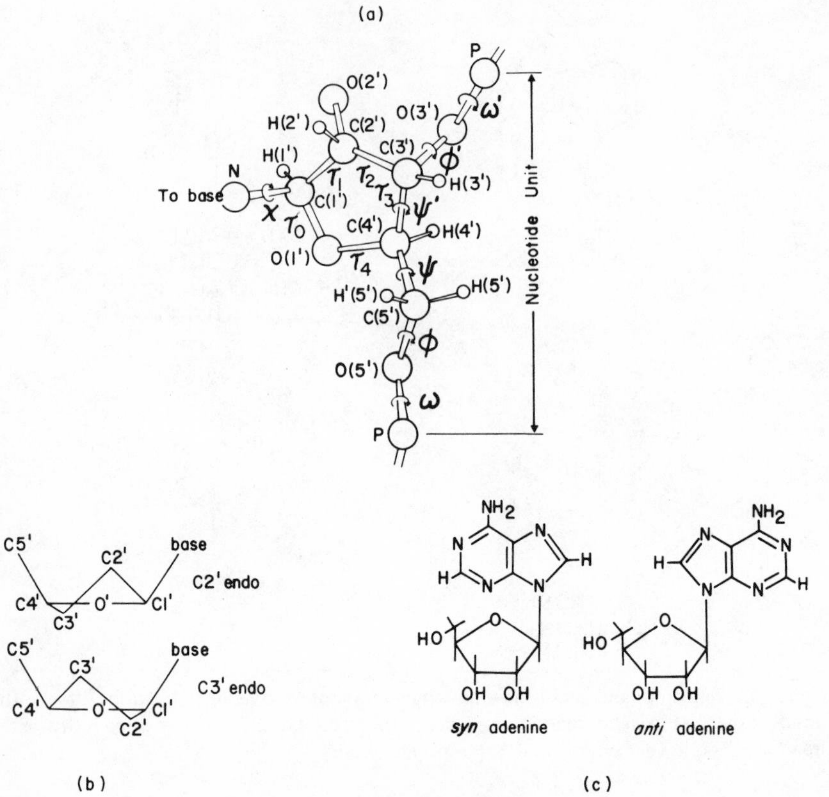

(a)

(b)   (c)

Figure 2 Nucleotide conformational features: (a) a nucleotide unit showing the torsional angles as Greek letters (9); (b) schematic illustration of the sugar ring puckering; (c) two adenosine conformations that differ in the angle (χ) about the glycosidic bond.

ATP refer to 5'-adenosine mono-, di-, and triphosphate, respectively; dApCp refers to 2'-deoxyadenylyl-(3'→5')-cytidine monophosphate.

Figure 2 displays a nucleotide unit together with its atom labeling scheme and the torsional angles within the unit. These torsional angles define the *conformation* of the nucleotide. An illustration of some conformations is also provided in Figure 2. The allowed and preferred conformations of nucleic acid constituents have been reviewed previously (9, 10). Ranges of torsional angles observed in crystal structures of nucleotides and from diffraction patterns of polynucleotides are given in Table 1. The definition of some of these angles warrants further discussion.

Unlike the nucleic acid bases, the ring atoms of a sugar do not lie in a common plane. Both X-ray crystallographic studies (11, 12) and NMR

Table 1   Preferred Conformational Angles of Nucleotides

Angle[a]	Range (degrees)	Conformation
χ	0–72	anti
	120, 220	syn
τ_0	-34–$+29$	
τ_4	-20–$+26$	
τ_1	314–346	C-3′ endo
τ_2	30–46	C-3′ endo
τ_3	316–338	C-3′ endo
τ_1	10–43	C-2′ endo
τ_2	320–343	C-2′ endo
τ_3	13–33	C-2′ endo
ω	273–295	Right-handed polynucleotide
	55–80	Left-handed polynucleotide
ϕ	125–220	
ψ	35–65	
ψ'	70–90	C-3′ endo
	140–160	C-2′ endo
ϕ'	150–270	
ω'	273–296	

[a]Definitions of angles are given in Figure 2. For definition of angle zeroes and further discussion, see references 9 and 10.

solution studies (13) reveal the sugar ring to be puckered. The particular conformation, or puckering, of the sugar ring may be defined in terms of the displacement of the carbon atoms, C2′ and C3′, from the plane defined by the atoms C-1′, C-4′, and O (Fig. 2b). Puckering displaces C-2′ and C-3′ above or below this plane. If the major displacement of an atom is toward the same side as the C-5′ carbon atom, the conformation is designated *endo*. Displacement toward the opposite side is an *exo* conformation. In a C-2′ endo conformation, for example, the C-2′ carbon is displaced ~0.5 Å from the ring plane on the same side as C-5′; and, correspondingly, the C-3′ carbon is displaced ~0.2 Å to the opposite side. This puckering of the sugar ring defines the stereochemistry of double-stranded polynucleotides, as will be seen later. Double-stranded DNA duplexes are characterized by both the C-2′ endo and C-3′ endo conformations, while in RNA duplexes only C-3′ endo conformations occur.

The orientation of the base relative to the sugar ring is another important conformational feature that determines the accessibility of base substituents to chemical attack. The angle of rotation, χ, about the glycosidic bond determines the relative orientation of the base with respect to the sugar ring (Fig. 2c). Values of χ less than 90°, where the exocyclic substituents of either purines or pyrimidines point away from the sugar ring, are referred to as *anti* conformations. The less common *syn* conformations correspond to values of χ greater than 90°. Pyrimidine nucleosides are almost exclusively in the anti conformation, the syn conformation being sterically hindered (14). Other interactions, such as hydrogen bonding in the case of 4-thiouridine (15) and 5'-methylammonium-5'-deoxyadenosine (16) or the steric interactions due to the bulky methyl group in 6-methyluridine (17) can stabilize the unusual syn relative to the anti conformation.

The preferred conformational angles about the phosphate ester linkages found in crystal structures of nucleotides serve as reasonable guides for the stereochemistry of the helical backbone of polynucleotides (9). Details of the geometry of nucleic acid polymers are discussed below.

2.1.2 Protonation Sites. Proton ionization equilibria for nucleic acid constituents provide a starting point for understanding sites of metal ion coordination. Nucleic acids have a large number of potentially reactive sites. While it has been relatively straightforward to obtain thermodynamic constants describing ionization equilibria, assignment of the sites of proton attachment has been more difficult. Conflicting results, based on differing experimental conditions such as solvent and ionic strength, have further complicated assignments. The literature in this area, which has been reviewed (11, 18), is replete with NMR, Raman, and electronic spectroscopic (19) as well as X-ray crystallographic determinations. A summary of pK values and protonation site assignments is provided in Table 2.

Atom N-1 is the most basic nitrogen of adenine and its derivatives,

Table 2   pK$_a$ Values and Protonation Sites of Base or Nucleoside

Compound	Ionization Site	pK$_a$
Adenine	HN-1$^+$	3.5–4.2
	HN-9	10
Guanosine	HN-7$^+$	1.9–2.1
	HN-1	9.0–9.2
Uracil	HN-1/HN-3	9.2–9.5
Cytosine	HN-3$^+$	4.0–4.2
Ribose	OH	12.3–12.5

having pK_a values ranging from 3.5 to 4.2. This assignment is supported by X-ray crystallographic studies of adenine hydrochloride (20), 5'-AMP (21), and 3'-AMP (22). Recent ^{15}N-NMR results (23), while definitively establishing primary protonation at N-1, reveal small chemical shifts in the N-6 and N-9 nitrogen resonances upon protonation. To explain this result the authors suggest either small amounts of protonation at these sites or distribution of the formal positive charge on N-1 to the exocyclic amino nitrogens by resonance effects. Proton ionization of neutral adenine (pK ~ 10) is generally thought to be at the N-9 position (24).

Guanine exists in solution as a mixture of tautomeric forms, with protonation occurring at either of the two imidazole nitrogen atoms, N-7 or N-9. In 9-substituted guanine derivatives, N-7 is the predominant site of protonation, with a pK_a of 1.9 to 2.1 (11, 18). ^{15}N-NMR studies (23), while supporting this assignment, also reveal protonation at N-3 in guanosine. Ionization of neutral guanosine (pK ~9) involves deprotonation of the amidate nitrogen atom N-1.

In neutral solution, uracil exists predominantly in the diketo tautomeric form. In basic solution there is a 1:1 mixture of forms deprotonated at N-1 and N-3 (25). The pK_a for this ionization is ~9.5. Proton NMR results (26) using ^{15}N-cytosine derivatives indicate the amidate nitrogen, N-3, to be the most basic nitrogen of cytosine and its derivatives, having a pK_a ~4.

Additional sites of deprotonation are possible for mononucleotides. The anionic character of poly- and mononucleotides is provided by the acid dissociation of the phosphate group, having pK_a ~1 (27). A second proton dissociation, occurring only in terminal phosphates, is characterized by a pK_a value of 6.0 to 6.4. The titration of this proton has been effectively monitored by Raman spectroscopy (19) in the cases of 5'-AMP and 5'-GMP. It appears that the presence of the negative charge on the phosphate group influences the pK_a values of the constituent bases. Comparative titrations of nucleosides and nucleotides show an increase in the acid dissociation constants of the base in the presence of the phosphate anion. In addition, each of the nucleotides has an ionizable sugar proton, pK_a 12.3 to 12.5.

2.1.3 Hydrogen Bonding and Stacking.

The helical structure of double-stranded polynucleotides is stabilized through hydrogen bonding interactions between bases and through the stacking of these resultant base pairs in a columnar array. The amidate and exocyclic amine nitrogen donors of one base are linked through hydrogen bonds to the keto oxygen and heterocyclic nitrogen acceptors of its pair. Some of the common base pairing schemes are given in Figure 3. As can be seen, in addition to

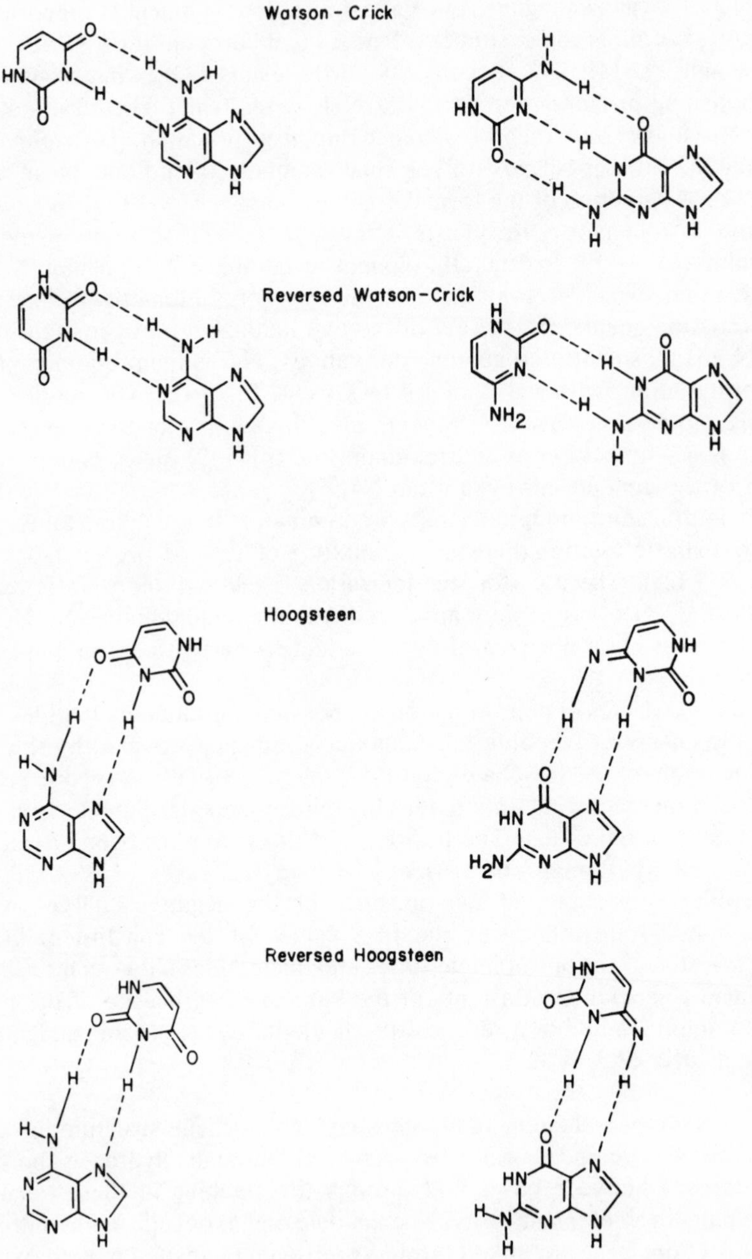

Figure 3 Various adenine–uracil (A–U), left, and guanine–cytosine (G–C), right, base pairing schemes.

Watson–Crick base pairing a number of other hydrogen bonding schemes are possible. These other pairing arrangements have been observed in the associations of nucleic acid constituents and moreover play a dominant role in stabilizing the structure of tRNA (vide infra).

In crystal structures of nucleic acid constituents, the bases are seen to form hydrogen bonds with themselves or with other bases (11). Proton NMR spectra of solutions reveal large chemical shifts for the resonances of protons involved in base pairing (28). Base pairing is less pronounced in nucleoside and nucleotide crystal structures than in those of the free bases. This may be the result of the steric bulk of the ribose ring or crystal packing influences. The sites of hydrogen bonding as well as the solvent and pH conditions that influence them must be taken into account in interpreting metal ion binding results. In a double-stranded polynu-cleotide, for example, only the N-7 and N-3 atoms of the purine rings are exposed, the remaining sites being involved in hydrogen bonding.

In addition to hydrogen bonding considerations, base stacking interac-tions provide a large degree of stabilization. The vertical stacking of purines and pyrimidines in the solid-state structures of mononucleotides resembles that found in polynucleotide structures (29). Usually base overlap is accomplished by positioning polar substituents over the ring system of a stacked base. These dipole-induced dipole forces also stabilize the interaction of metal complexes having planar heterocyclic rings as ligands with nucleic acids or their constituents. As will be seen below in solution as well as in the solid state, a noncovalent stacking interaction of the type described may predominate over covalent binding lacking this feature.

2.2 Fundamentals of Heavy Metal Binding

Here we examine the primary binding modes of heavy metals to mono-mers, namely, the interactions with the phosphate, sugar, and base moieties of the nucleotide. The bases afford the most number of sites for metal ion reactivity. Their heterocyclic and amidate nitrogen atoms are apt to form covalent bonds to the soft heavy metal centers. In contrast, few com-plexes are observed between the nonpolarizable ribose ring oxygen atoms and a soft metal cation. While the purines and pyrimidines serve as excellent ligands for metal coordination, the stability constants of metal–nucleotide complexes have a large component of electrostatic attraction between the electrophilic metal center and the anionic phosphate group. Solution studies indicate that transition metal ions have a high affinity for phosphate coordination, the bases playing a secondary role. The large number of crystal structure determinations of metal–base complexes, which are easier to crystallize than metal–nucleotide complexes, has shifted attention away from the phosphate group as a primary site of heavy metal binding. Specificity of metal binding to phosphates may be achieved by conformational effects. As we have seen, the stereochemistry of a nucleotide is determined by the torsional angles about the phosphate–sugar ester linkages.

2.2.1 Phosphate Interactions. In living cells, divalent metal ions coordinated to the phosphate groups of nucleotides, promote phosphate ester hydrolysis, and maintain polynucleotide structure. The complex between magnesium ion and ATP is a necessary cofactor for phosphoryl and nucleotidyl transfer enzymes as well as a major energy storage depot for the cell. It has been the subject of much investigation (30–32). How does the magnesium ion influence the structure of ATP and determine its selectivity for enzyme activation?

Divalent metal ions of the alkaline earth series bind exclusively to the phosphate groups of nucleotides. The stability constants increase for A $<$ AMP $<$ ADP $<$ ATP, reflecting the increasing phosphate content, and hence negative charge, of the ligand (33). There is in addition a decrease in stability with increase in atomic number down the group, as expected from charge-to-radius ratio considerations. These two trends combine to make magnesium–ATP a very stable complex.

Paramagnetic lanthanide ions have been used as NMR probes of the conformation of metal coordinated ATP in solution (34). Pseudocontact shifts of both ^1H and ^{31}P resonances of ATP were determined as a function of bound Ln^{3+} (Pr, Nd, Eu, Yb) concentration. Calculations of metal–proton and metal–phosphorus distances based on the chemical shift changes led to models of the Ln^{3+}–ATP structure with an anti arrangement of the base ribose unit and a helical phosphate chain like that observed in crystal structures of the triphosphates. The lanthanide ion is found to bind predominantly to the β and γ phosphates, coordinating to one β and two γ oxygen atoms. No interaction is observed with the purine ring. Given the absence of purine interactions, the Ln^{3+}–ATP complex appears as an interesting chemical approximation for the biologically essential Mg^{2+}–ATP complex.

Alkaline earth and lanthanide cations are not the only ones that form stable links to the phosphate groups. Potentiometric and spectrophotometric data indicate that the primary stabilization of transition metal–nucleotide complexes is through phosphate interactions, the nature of the purine or pyrimidine base having a comparably small effect on complex stability constants (35, 36). As one moves across and down the periodic table, covalent binding to the bases becomes more important

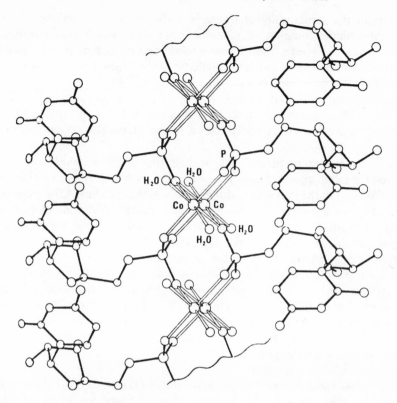

Figure 4 The polymeric chain in $[Co_2(H_2O)_4(5'-UMP)_2]_n$, which is propagated by phosphate groups bridging neighboring metal atoms. [Reproduced with permission from *Biochim. Biophys. Acta,* **477**, 197(1977). Copyright by Elsevier/North-Holland Biomedical Press.]

than phosphate binding, consistent with the increased softness of the metal. The 40% higher association constant of Mn^{2+} over Mg^{2+} for 5'-GMP, obtained through an ESR-monitored competitive binding experiment (37), is attributed to base binding. Under moderately high ionic strength conditions (0.1 M $NaClO_4$), no interaction is seen between the phosphate moiety and methylmercuric ion, as determined by Raman difference spectroscopy (38).

Covalent binding to phosphate also occurs with metals to the right side of the transition series and should never be ruled out without cause. In both $[Cu(3'-GMP)(phen)(H_2O)]_2$ (39) and $[Co_2(5'-UMP)_2(H_2O)_4]_n$ (40), the metal binds exclusively to the phosphate moiety. In the latter case, a polymeric chain is propagated through alternating metal–(phosphate oxygen)–metal binding (Fig. 4). The covalent metal–pyrophosphate bonds in the structurally characterized dimeric cis-diammineplatinum–pyrophosphate complex (41) demonstrate that strong association can occur even between a "soft" heavy metal and phosphate oxygen atoms. In this instance, the pyrophosphate anion bridges two cis-diammine-platinum(II) units through four metal–oxygen linkages. These examples serve as reminders that, in interactions of most metal ions with nucleic acids, strong electrostatic interactions occur with the phosphate groups and often involve direct coordination to the phosphate oxygen atoms.

2.2.2 Ribose Interactions. The pentose ring of the nucleotide is a poor ligand for metal coordination. Few examples of metal–ribose interactions are reported in the literature. Magnetic resonance studies (42) support the formation of a copper–nucleotide chelate, under conditions of high pH (pH 8.5 to 10), through the 2'- and 3'-hydroxyl groups of the ribose. Under these conditions copper ion can discriminate between ribo and deoxyribo derivatives. Osmate esters of the ribose unit are most commonly cited as examples of heavy metal–ribose binding. In the structure of $Os(adenosyl)O_2(py)_2$ (43), adenosine is bound to the osmate anion through diester linkages involving the 2'- and 3'-hydroxyl groups. The reaction of the osmate anion with the cis-2',3'-diol group of the ribose ring has been suggested for heavy metal staining of RNA in electron-microscopic investigations.

2.2.3 Base Interactions. In an early study (44) designed to establish the sites and association constants for binding mercuric ion to nucleosides, spectrophotometric titrations of the constituent bases as a function of pH and metal concentration were carried out. The spectra revealed partial mercuration at one or more sites. The sites of metallation were assumed to be identical with the sites of proton displacement. Subsequent spec-trophotometric studies (45–47) of a series of metal–nucleoside complexes employed alkylated derivatives of the bases in order to establish the metal binding sites. Proton and ^{13}C-NMR studies (48–50) monitoring chemical shift perturbations and line broadening as a function of metallation, and Raman difference spectral measurements (51, 52) following changes in vibrational modes on metallation have provided additional data support-ing the assignments of metal binding sites. Table 3 shows the ^{13}C chemical shifts of the nucleosides in the presence and absence of (en)Pd(II) (48). Metallation at the N-7 nitrogen in purines causes a substantial downfield shift of the neighboring C-8 resonance. In pyrimidines, metal binding through N-3 results in pronounced shifts of the C-2 carbon atom reso-nance. These perturbations can be used to establish the primary sites for metallation of the bases.

Table 3 ¹³C Chemical Shifts[a] of Nucleosides in D_2O Upon Protonation and Metalation[b]

Pyrimidine	C-4	C-2	C-6	C-5	C-1'	C-4'	C-2'	C-3'	C-5'	CH₃
Cytidine										
HL⁺	92.9	82.0	77.9	28.8	24.0	18.0	7.7	2.8	-5.9	
L	99.1	90.5	74.7	29.2	23.4	16.9	7.2	2.4	-6.1	
(en)PdL₂²⁺	98.7	88.0	75.2	29.3	23.8	17.4	7.2	2.4	-6.1	
Thymidine										
L	99.5	84.8	70.6	44.6	18.4	19.8	-28.2	3.7	-5.5	-55.2
L⁻	109.6	92.6	69.5	44.9	18.4	19.4	-28.0	4.0	-5.2	-53.9
(en)PdL₂	106.8	90.0	68.5	43.5	18.5	19.2	-28.5	3.6	-5.6	-54.2
Uridine										
L	99.2	84.6	74.8	35.2	22.4	17.2	6.8	2.5	-6.2	
L⁻	110.0	92.7	73.7	36.1	23.1	17.1	6.8	3.0	-5.5	
(en)PdL₂	107.0	89.9	72.9	34.7	23.2	16.9	6.6	2.6	-6.1	

Purine	C-6	C-2	C-4	C-8	C-5	C-1'	C-4'	C-2'	C-3'	C-5'	CH₃
1-CH₃ Guanosine											
L⁻	90.8	87.2	81.7	70.8	48.8	20.9	18.3	6.6	3.6	-5.4	-38.4
(en)PdL₂²⁺	89.8	88.3	81.6	72.3	47.4	21.5	18.6	7.0	3.2	-5.8	-38.2
Guanosine anion											
L⁻	100.8	93.7	83.9	69.6	51.6	21.2	18.9	6.2	4.2	-5.1	
(dmen)PdL₂	98.5	92.0	83.9	70.4	50.3	21.3	18.7	6.2	3.9	-5.2	

[a]In ppm downfield from external dioxane.
[b]Table reproduced with permission from reference 48.

45

Table 4   Metalation Sites on the Nucleosides

Compound	Common Coordination Site	Additional Site
Adenosine	N-1	N-1
Guanosine	N-7	O-6, N-1
Uridine	N-3	O-2, O-4
Cytidine	N-3	N-2, O-4

[handwritten annotations: "pg 35", "N-4, O-2"]

   The primary heavy metal binding sites are given in Table 4. X-Ray
crystallographic determinations (2) support these assignments. Sites of
protonation serve as preliminary guides for the centers of metal reactivity.
The relative softness, or polarizability, of the ligating base as well as pH
and solvent conditions are the prime factors governing the binding of
heavy metals to the base substituents. The sulfur atoms of thiolated bases
are especially reactive for heavy metal coordination, forming stable complexes.
   Various authors (52, 53) have emphasized the importance of pH in
determining metalation sites. ^{13}C-NMR results (53) show that Pt(II) binds
N-1 and N-7 of adenosine almost equally at pH 5. At lower pH, where N-1
becomes protonated (pK_a 3.6), the availability of that site for coordination
is diminished. For both adenosine and guanosine, N-7 and N-1 are equally
favored in neutral solution, while N-7 is the only binding site under acidic
conditions. The greater reactivity of N-7 is also a consequence of the fact
that N-1 better participates in favorable base pair hydrogen bonding
interactions (cf. Watson–Crick base pairing in polynucleotides). Metalation of pyrimidines is almost exclusively at N-3. Here again, in the case of
uridine and thymidine, there is only sparse metal reactivity under acidic
conditions, a high fraction of molecules being protonated at N-3. At
neutral pH, reactivity at this site is favored.
   Metalation at one site on a base tends to affect the reactivity of other
sites toward protonation (54) and metal substitution (45). Binding of
pentammineruthenium(III) to N-7 of guanine decreases the pK_a values of
both N-9 and N-1 by two to three units. An adenosine complex with
platinum(II) atoms bound both to N-7 and N-1 of the same base has been
isolated and structurally characterized (55). Thus the binding of a metal
ion at one site has a marked influence on the acidity of the base moiety
and, under conditions of relatively high (1:1) metal-to-base ratios, will
lead to additional substitution reactions.
   Another site of base reactivity is the electron-dense C-5–C-6 double

bond in pyrimidines. By means of a simple acetoxymercuration reaction, mercury may be covalently attached to the C-5 atom of UTP or CTP (56). The acidic conditions used in these reactions favor metal substitution at the C-5 carbon atom rather than coordination to a ring nitrogen. Derivatives of guanosine and adenosine, presumed to be mercurated at the C-8 carbon atom, have been similarly prepared but were not fully characterized. The crystal structure (57) of a ruthenium(III)–caffeine complex, isolated in dilute acid, does reveal metalation at C-8. The caffeine ligand, a pyrimidine base alkylated at N-1, N-3, and N-7, presents considerable steric hindrance to metalation at the usually favored imidazole nitrogen, N-9, leading to the formation of a carbon-bound metal complex.

Reactions of bis(pyridine)osmium tetroxide with pyrimidines are also directed to the C-5–C-6 center but do not result in the formation of a metal–carbon bond. Addition across the C-5–C-6 double bond produces an osmate ester of thymine (58, 59). The reaction product is analogous to that formed in the reaction of osmium tetroxide with the *cis*-2′, 3′diol of the ribose ring. Both reactions are depicted in Figure 5. The pyrimidine osmate esters and mercurated pyrimidines are relatively stable heavy metal complexes and have been utilized extensively in electron microscopic investigations. They will be considered in that context later.

The relative reactivity of the bases toward a given heavy metal ion is another question of interest. The substitutionally inert third row transition metal ions are under kinetic control in their reactivity with the bases. The order of nucleophilicity of the nucleotides, GMP > AMP >> CMP >> UMP (52), parallels the relative reaction rates of cis-$[(NH_3)_2Pt(OH_2)_2]^{2+}$ with the bases (60). Equilibrium constants for the different reactions, on the other hand, are of comparable magnitude. Experiments testing comparative base reactivity toward CH_3HgOH, a substitutionally more labile metal complex, reveal metal binding with deprotonation at N-3 of TMP to be most favored. By contrast, thymine is the least reactive base for platination. The relative rapidity of reactions under kinetic control thus leads to differences in final product formation. This feature can be used to achieve selective metallation of nucleotides, a subject that is more fully discussed in the following section.

2.3 Selective Heavy Metal Binding Modes

Given the general scheme of metal coordination to the amidate and heterocyclic ring nitrogen atoms of the bases, how might selectivity in metal–nucleic acid interactions be achieved? We have already seen that conformation can determine selectivity in the case of metal–ATP binding. This factor will be shown to be an important source of reaction specificity

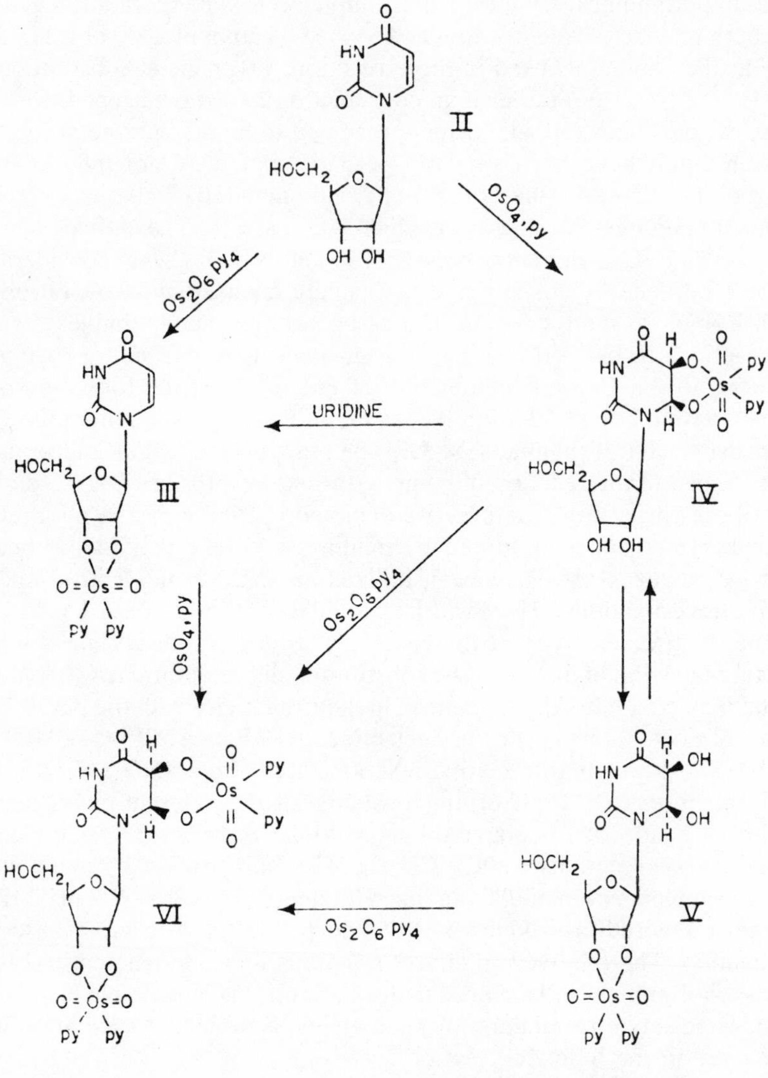

Figure 5 Reaction pathways of uridine with oxoosmium–pyridine complexes. (Reproduced with permission from *J. Am. Chem. Soc.*, **97**, 7352 (1975). Copyright by the American Chemical Society.)

when we examine metal–polynucleotide interactions. Also mentioned above were the effects of pH and relative reaction rates as factors leading to selective differences in product formation. Three important sources of metal binding specificity have not yet been addressed. One is the exocyclic substituents of the bases. The ability of these substituents to determine specific complex formation, both through direct bonding interactions and through nonbonding effects, is being increasingly recognized (61). A second source of bonding specificity is stacking interactions that occur when the metal complex has planar, aromatic ligands (62). Below we consider some intercalation complexes, where stacking interactions result in unusual complex geometries. Finally, specific binding modes in metal–nucleotide complexes can result when the metal coordinates to a combination of base, sugar, and phosphate moieties, leading to an extensive polymeric array (6). These three sources of metal binding specificity will now be illustrated. Since metal–nucleotide complexes are difficult to crystallize, few detailed structural studies are available in the literature. We therefore include in our discussion some first-row transition metal–nucleotide complexes. The interaction of transition metals with mononucleotides in many respects serves as a model for the binding of heavy metals to a polynucleotide.

2.3.1 Exocyclic Base Substituents. Perhaps the most effective way of achieving reaction specificity is to use thio-substituted bases. The soft polarizable sulfur atom serves as a highly reactive center for preferential heavy metal coordination. ^{13}C-NMR measurements (63) of dimethyl sulfoxide-d^6 solutions containing various thiolated and nonthiolated nucleosides in the presence and absence of excess mercury reveal that both CH_3HgCl and $HgCl_2$ bind principally to the sulfur atom of s^6-guanosine and s^8-guanosine. As shown in Figure 6, there is a dramatic 15.1 ppm upfield shift in the C-6 resonance upon mercuration at sulfur atom S-6. The structure (64) of bis(6-mercapto-9-benzylpurine)palladium(II) (Fig. 7) exemplifies metal chelation by a thiolated purine base. Here two 6-mercaptopurine ligands are coordinated to the metal center in a cis orientation through the heterocyclic ring nitrogen atom N-7 and the thiolato sulfur atom S-6, forming two five-membered chelate rings. The great affinity of soft heavy metal atoms for sulfur donors determines this unusual metal–base complex geometry. A primary target of metalation of certain *E. coli* tRNAs is the 4-thio-U base, due to the high association of heavy metals with sulfur. Indeed, thiolated nucleotides generally have proved to be highly specific centers for stable heavy metal attachment.

The exocyclic oxygen atoms of both purine and pyrimidine bases also participate in selective metal reactions. While the affinity of heavy metals

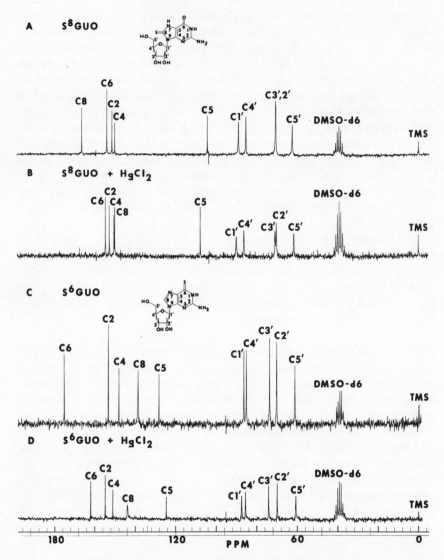

Figure 6 ^{13}C-NMR spectra and assignments of (A) 0.5 M s^8-guanosine; (B) 0.5 M s^8-guanosine plus 2.0 M HgCl$_2$; (C) 0.5 M s^6-guanosine; and (D) 0.5 M s^6-guanosine plus 2.0 M HgCl$_2$ in dimethyl sulfoxide (DMSO). The C2' and C3' assignments in panels C and D should be interchanged (see *Biochem. Biophys. Res. Commun.,* **46,** 808 (1972)). [Reproduced with permission from *Biochim. Biophys. Acta,* **402,** 403 (1975). Copyright by Elsevier/North-Holland Biomedical Press.]

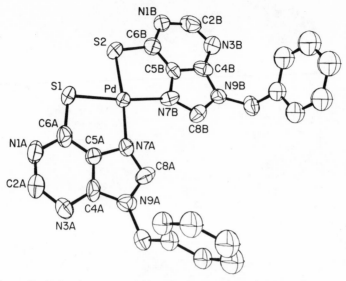

Figure 7   Molecular structure of bis(6-mercapto-9-benzylpurine)palladium(II).

for a hard, electronegative oxygen atom does not match that for sulfur or even the nitrogen atoms of the bases, it has become quite clear that exocyclic oxygen substituents direct specific metal interactions. A prime example is that of the platinum pyrimidine blues (65–67). These anomalously colored complexes, formed on incubation of *cis*-DDP with any of a number of substituted pyrimidines, are thought to be mixed valent polymers, linked through fractional metal–metal bonds and amidate bridges. The platinum pyrimidine blues have thus far been isolated only as amorphous powders, presumably because of their polymeric nature. The structure (68) of an analog of the platinum pyrimidine blues, *cis*-diammineplatinum α-pyridone blue, is shown in Figure 8. While this complex, unlike all others discussed thus far, is not strictly speaking a metal–nucleic acid complex, we believe its structure to embody the characteristic features of platinum pyrimidine blues. The analog of uracil, α-pyridone, used in the synthesis to prevent the formation of mixtures differing in the site of metal attachment to the ligand, has a single cyclic amidate group as compared to the two centers, N-1–O-2 and N-3–O-4, that occur in uracil (Fig. 9). X-Ray photoelectron spectra (69) reveal the similarity in electronic structure of *cis*-diammineplatinum α-pyridone blue and other platinum blues. Solution chemical and spectroscopic studies (70) also demonstrate that the α-pyridone blue is quite similar to its pyrimidine analog, *cis*-diammineplatinum uracil blue. As can be seen in

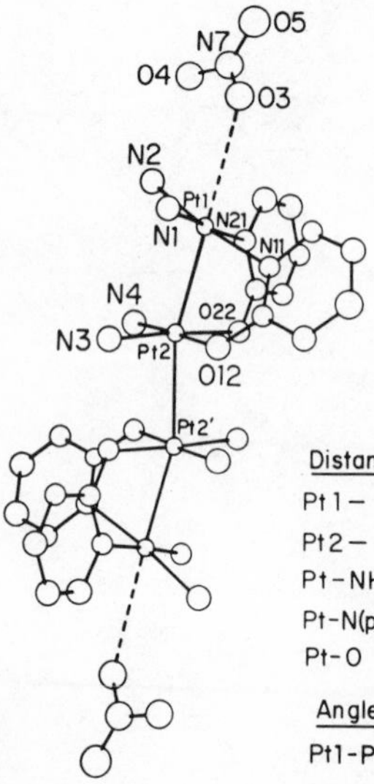

Distance	Å
Pt 1 – Pt 2	2.77
Pt 2 – Pt 2'	2.88
Pt – NH₃(av)	2.03
Pt – N(pyridone)	2.03
Pt – O	2.02

Angle	Deg.
Pt1–Pt2–Pt2'	164.6°

Figure 8 *cis*-Diammineplatinum α-pyridone blue.

Figure 8, the partially oxidized tetranuclear cation is actually a dimer of dimers, linked to one another by a metal–metal bond and interligand hydrogen bonding between the ammine nitrogen atoms and pyridonate oxygen atoms. Each dimer is composed of two *cis*-diammineplatinum units bridged by two α-pyridonate ligands through the exocyclic oxygen and deprotonated nitrogen atoms. An Extended X-Ray Absorption Fine Structure (EXAFS) study (71) reveals platinum–platinum bonds in the uracil blue comparable in length (2.9 Å) to the distances observed in the tetranuclear chain of the pyridone blue. The polymeric structure of *cis*-diammineplatinum uracil blue may be propagated using alternate singly bridged uracilate anions, with platinum coordination to a mixture of N-1 and O-2 or N-3 and O-4 ligating atoms.

The structure (72) of a *cis*-diammineplatinum 1-methylthyminate dimer reported recently closely resembles that of the α-pyridone blue. As seen

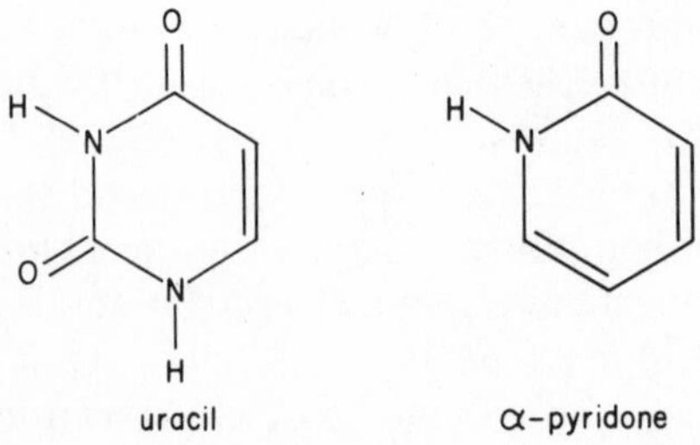

uracil

α–pyridone

Figure 9 Structural formulae of uracil and α-pyridone.

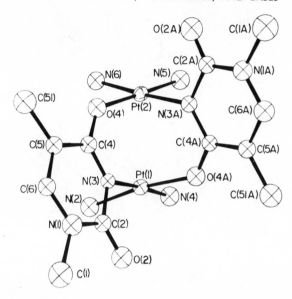

Figure 10 Structure of the dimeric *cis*-diammineplatinum(II) 1-methylthyminate dication. [Reproduced with permission from *J. Am. Chem. Soc.*, **100**, 3371 (1978). Copyright by the American Chemical Society.]

in Figure 10, two 1-methylthyminate ligands bridge *cis*-diammineplatinum units through the heterocyclic ring nitrogen atoms, N-3, and exocyclic oxygen atoms, O-4. Unlike the α-pyridone blue, the amidate anions bridge platinum centers in a head-to-tail fashion. As expected, the metal–metal distance in this platinum(II) species is longer than in the partially oxidized α-pyridonate analog, which has a formal oxidation state of 2.25. As in the α-pyridone blue structure, the exocyclic substituents serve to bridge metal centers forming a discrete oligomeric structure.

The dimeric complex of 1-methylcytosine with silver (73) also exemplifies the coordination of an exocyclic oxygen atom to soft transition metal centers. As shown in Figure 11, cytosine anions bridge the tetrahedral silver cations through the heterocyclic N-3 nitrogen and exocyclic O-4 atoms. No short-range metal–metal interaction is observed. The isolation (70) of a *cis*-diammineplatinum hypoxanthine green suggests that in purine complexes as well exocyclic oxygen atoms may participate in metal coordination.

Examples of metal binding by exocyclic substituents are not restricted to metal–oxygen complexes. A recently isolated *cis*-diammineplatinum(2.5) cytosinate dimer (74) reveals metal coordination through the heterocyclic N-3 nitrogen and deprotonated exocyclic amine (N-4)

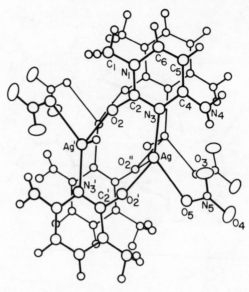

Figure 11 Structure of the nitrate salt of the dimeric 1-methylcytosine silver(I) complex. [Reproduced with permission from *J. Am. Chem. Soc.*, **99**, 2797 (1977). Copyright by the American Chemical Society.]

atoms. High pH conditions were presumably required to form this product. While a deprotonated exocyclic amine serves as an excellent ligand for soft heavy metal coordination, it is unlikely that this mode of binding is representative of reaction products in neutral aqueous solution. The various examples discussed above demonstrate however the important participation of exocyclic substituents in direct metal–ligand bond formation. In all instances cited, the nucleic acid bases serve to bridge two metal centers through ring nitrogen atoms and appropriate exocyclic substituents, forming very specific, compact structures. In view of the close proximity of one base to another on the polynucleotide chain, these structures may point to features of metal–polynucleotide interactions.

Exocyclic substituents also appear to influence the stereochemistry of metal–nucleic acid complexes through weaker, secondary interactions, if not through direct coordination. A characteristic feature of copper cytosine complexes (61), apart from metal coordination at the N-3 nitrogen atom, is a weak axial interaction at the O-2 oxygen atom. In the structure (75) of the cytidine complex of glycylglycinatocopper(II), the coordination about the copper is square pyramidal, with the dianion of glycylglycine and N-3 of cytidine occupying the four equatorial sites in the coordination sphere and the O-2 atom of cytidine at the axial site. The copper-oxygen bond distance is 2.74 Å. The metal–purine complex, (N-3, 4-benzosalicylidine-N',N'-dimethylethylenediamine)(theophyllinato)-copper(II) (76), again reveals a square pyramidal geometry about the copper. Here the Schiff's base chelate ligand and the N-7 atom of the theophylline anion occupy the four equatorial positions. The O6 oxygen atom of the theophylline ligand is weakly bonded at the axial site, with a Cu–O distance of 2.92 Å. The authors suggest that steric crowding and the absence of interligand hydrogen bonding promotes this weak bidentate metal–purine interaction. The mercury adducts (77) of uracil and hydrouracil reveal an unusual mode of coordination. In these complexes the $HgCl_2$ unit is bound not to the N-3 nitrogen atom but instead through long Hg–O bonds (2.71 and 2.88 Å) to the exocyclic oxygen atom O4.

The exocyclic keto and amino groups of the bases frequently act as hydrogen bond acceptors and donors, respectively, for other ligands coordinated to the metal. The complex (glycylglycinato)(aquo)(9-methyladenine)copper(II) (78) has square pyramidal geometry and metal coordination to the base through N-7. In addition, the exocyclic amine of adenine forms a strong interligand hydrogen bond to an axially coordinated water molecule. Intramolecular hydrogen bonding also occurs in the isostructural complexes $[Pt(Guo)_2(en)]^{2+}$ (79) (Fig. 12) and $[Pt(Guo)_2(NH_3)_2]^{2+}$ (80). The interaction occurs between the amine ligands of one molecule and carbonyl oxygen of an adjacent guanine moiety. The metal atom is bound to each base through the N-7 ring nitrogen atom. An additional noteworthy feature of these complexes is the stacking of guanine bases of adjacent molecules at a distance of 3.35 Å.

2.3.2 Stacking Interactions. Stacking interactions also affect the formation of metal–nucleotide complexes. Planar heterocyclic ligands coordinated to the metal can result in specific complex geometries. In the unusual stacked structure of the 2:2 chloroterpyridineplatinum (II):adenosine 5'-monophosphate complex (81) there is no direct metal–purine coordination. Rather, a hydrogen-bonded AMP base pair is intercalated between two chloroterpyridineplatinum(II) cations. The base pair, a Watson–Crick/Hoogsteen hybrid, is formed between the donor exocyclic amines, N-6, and acceptors N-7 and N-1. The geometry of the complex maximizes the overlap of the terpyridine ligands with the adenine rings. Platinum complexes with planar aromatic heterocyclic ligands have been shown to bind noncovalently to DNA by intercalation (see Section 3.2.2). The platinum complex cited above and the complex (82) of $[(terpy)Pt(HET)]^+$ with the dinucleoside monophosphate, dCpG, shown in Figure 13 exemplify this interaction mode. Stacking forces stabilize the substitutionally inert terpyridineethanethiolatoplatinum(II)

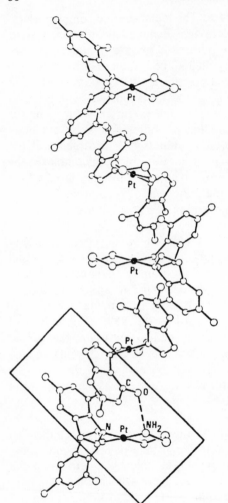

Figure 12 A fragment of the spiral consisting of [Pt(en)(guanosine)$_2$]$^{2+}$ units. Boxed area shows two stacked guanine rings from neighboring units; the (en)Pt(II) moiety links these bases through coordination to the N-7 nitrogen of one base and hydrogen bonding between the amine ligand and the O-6 oxygen atom of the other. Ribose rings have been omitted. [Reproduced with permission from *J. Am. Chem. Soc.*, **97**, 7379 (1975). Copyright by the American Chemical Society.]

complex with the dinucleoside monophosphate dCpG. The ring nitrogen atoms of the tightly chelating terpyridine ligand and the sulfur atom of the ethanethiolato anion comprise the primary coordination sphere. Guanine–cytosine base pairs are stacked above and below the platinum coordination plane at a distance of 3.4 Å. The resultant stereochemistry resembles that of an intercalated fragment of a double-helical DNA polymer. The sugar rings adopt different puckering angles on either side of the intercalator, having a C-3′ endo conformation at the 5′-end of the dinucleoside phosphate and the normal C-2′ endo conformation at the 3′-end. The metal coordinated terpyridine ligand is thus seen to stabilize a

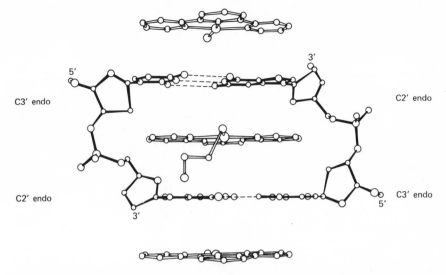

Figure 13 View of the double-helical dinucleoside monophosphate fragment, dCpG, with [(terpy)Pt(HET)]⁺ intercalated between, as well as stacked above and below, the base pairs. The hydroxyethyl side chain is shown only for the intercalated complex. (Reproduced with permission from ref. 82.)

specific binding mode. As shown in Section 3.2, additional specificity can be achieved by a noncovalent, intercalative stacking interaction through the preferential association of metallointercalating reagents with a particular base pair site.

2.3.3 Metal–Nucleotide Polymers. Hydrogen bonding and stacking interactions often stabilize polymeric structures in the crystal lattice. It is therefore not surprising that polymeric arrays predominate in structures of metal complexes of nucleoside phosphates. The adducts [Pt(en)(5'-CMP)]₂ (83) and [Zn(5'-CMP)]ₙ (84) illustrate this bonding pattern. In both complexes the 5'-CMP ligand serves to bridge the metal centers. In the former binuclear complex, an ethylenediamine ligand, the N-3 nitrogen atom of one cytosine base and a phosphate oxygen atom of the other cytosine monophosphate ligand comprise the primary square planar coordination sphere about each platinum atom. A strong platinum to phosphate oxygen linkage at a distance of 1.97 Å is a noteworthy feature of this structure. In the latter compound each tetrahedral zinc atom is coordinated to the N-3 nitrogen atom of one cytosine base, two phosphate oxygen atoms of different CMP ligands, and a water molecule. In addition, there is a weak interaction with the exocyclic O-2 oxygen of the cytosine base that is primarily coordinated through its ring nitrogen. Three different CMP ligands are bound to each metal center, and three

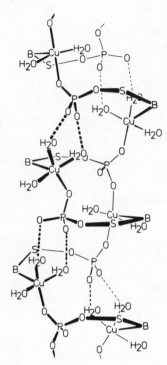

Figure 14 Schematic illustration of the polymeric structure of $[Cu_2(5'\text{-GMP})_3(8H_2O) \cdot 4H_2O]_n$. The sugar and base groups are depicted as S and B, respectively. [Reproduced with permission from *Biochim. Biophys. Acta*, **425**, 369 (1976). Copyright by Elsevier/North-Holland Biomedical Press.]

metal atoms are coordinated to each CMP. The result is a complex, crosslinked polymeric structure.

The complex (85) $Cd_2(5'\text{-IMP})_3 \cdot 12H_2O$ exhibits the full range of nucleotide binding modes. One of two crystallographically independent cadmium atoms is coordinated to the purine ring at the N-7 position and two oxygen atoms (O-2' and O-3') of a ribose group. The other is bound to two purine bases through N-7 nitrogen atoms and to a phosphate oxygen atom. Water molecules complete the octahedral coordination positions of each metal. Finally, the complex (86) $[Cu_3(5'\text{-GMP})_3(H_2O)_8 \cdot 4H_2O]_n$, schematically shown in Figure 14, generates a helical polymer stabilized by phosphate bridges and a network of hydrogen bonds. Each copper atom, having square pyramidal coordination geometry, is bound to phosphate and water oxygen atoms at the pyramid base and axially to the purine heterocyclic nitrogen atom, N-7. Base stacking occurs between purines of adjacent polymer units, which generates a columnar array of stacked purine rings.

Features of the coordination geometry and nucleotide conformational angles of the complexes discussed above are given in Table 5. These

Table 5 Coordination Geometries and Conformations of Various Heavy Metal–Nucleic Acid Complexes[a]

Complex	Geometry About Metal	Coordination Site	Sugar Puckering	χ_{C-N}	Reference
Os(Ado)(py)$_2$O$_2$	Octahedral	O2' O3'	C-2' endo	syn	43
Pt(Guo)$_2$(en)$^{2+}$	Square planar	N7	C-3' endo	anti	79
cis-Pt(Guo)$_2$(NH$_3$)$_2^{2+}$	Square planar	N7	C-3' endo	anti	80
[(terpy)PtCl](5'-AMP)	Square planar	Stacked	C-2' endo C-4' exo	anti anti	81
[(terpy)Pt(HET)](dCpG)	Square planar	Stacked	C-2' endo C-3' endo	anti anti	82
[Pt(en)(5'-CMP)]$_2$	Square planar	N3 O(phos)	C-2' endo C-2' endo	anti anti	83
[Zn(5'-CMP)(H$_2$O)]$_n$	Tetrahedral	N3 O(phos)	C-2' endo	anti	84
[Cd$_2$(5'-IMP)$_3$(H$_2$O)$_6$]$_n$	Octahedral	N7 O(phos) O-2' O-3'	n.a.	n.a.	85
[Cu$_3$(5'-GMP)$_3$(H$_2$O)$_8$]$_n$	Square planar	N-7 O(phos)	C-3' endo C-3' endo C-2' endo	anti anti anti	86

[a] Adapted from reference 6; n.a. means not available.

examples illustrate the rich variety of binding modes and common characteristics of metal–nucleotide complexes. Coordination through the primary metal binding sites of the purines and pyrimidines is usually observed, although additional phosphate coordination distinguishes the geometry of these complexes from that of metal–base adducts. The wealth of available sites on a nucleotide for metal coordination leads to a ubiquity of polymeric structures. Hydrogen bonding and stacking interactions further stabilize the intricate polymeric geometries. The conformational angles about the sugar and phosphate linkages are within the range found in polynucleotides. Structural distortions may arise so as to maximize metal coordination or to relieve crystal packing constraints.

3 METAL BINDING TO POLYNUCLEOTIDES

3.1 Polynucleotide Structure

The structure of the DNA double helix is familiar to many. As proposed by Watson and Crick (87) DNA is a right-handed double helix. The sugar phosphate backbone on the outside minimizes electrostatic repulsions between the anionic groups. The paired bases form a stacked column within the helix. The double-helical backbone generates ribbons of polar substituents about the axis, the deep grooves exposing base substituents to solution. Although alternative models for DNA have been proposed (88), the Watson–Crick structure is still the most widely accepted one. In this section we examine some aspects of the secondary structure of the helix as well as some examples of DNAs having more complex tertiary structure than that of the linear duplex.

The long, flexible, polymeric nature of the double helix allows for its characterization in solution by physical methods that measure its distinctive shape. The basis for many of these techniques is well described in a number of texts (89–91). It is important however to consider for a moment the variable long-range structure of a nucleic acid polymer. The viscosity and sedimentation properties of a polymer, as well as its electrophoretic mobility in a gel, depend not only on its molecular weight but also on its frictional coefficient. The flexible, randomly coiled, looped structures of denatured, single-stranded polynucleotides sediment far more slowly, for example, than the stiff double-stranded counterpart. The binding of metal ions can lead either to a compact, highly coiled structure of the nucleic acid or to an extended flexible shape. The melting temperature, as determined by the changes with temperature in hyperchromicity of the absorption spectra of a polynucleotide, also reflects the polymer structure in solution. Melting of the double helix, that is, its complete separation into single strands, occurs at low temperatures whenever the helix is destabilized. Helix destabilization is characterized by unstacking of the bases and cleavage of the base-pair hydrogen bonds. As we see below, changes in the physical properties of polynucleotides in the presence of metal ions demonstrate the marked effects of metal binding on their tertiary structure.

3.1.1 DNA and RNA Duplexes. Watson–Crick base-paired duplexes can assume a number of unique secondary structures (10). The helical conformations, or forms, of RNA and DNA are characterized by the torsional angles about the phosphate and sugar linkages. They can be defined in terms of a few parameters of the helix. The axial translation, or rise per residue, corresponds to the distance along the helix axis between neighboring base pairs. The rotation per residue is the angular displacement between base pairs about the helix. The dimensions of a helix may be alternatively described by the helix pitch, the axial distance for one turn of the helix, and the total number of base-paired residues per turn. The tilt, or angle subtended by the plane perpendicular to the helix axis and that of a base pair, describes another feature of the duplex geometry. While values for these parameters vary somewhat with the polynucleotide form, secondary structures fall primarily into two main classes, designated as the A and B forms of DNA.

The various polynucleotide conformations, reviewed (92) recently, have been discovered chiefly through X-ray diffraction analysis of oriented, polycrystalline fibers of DNA (93–95) and RNA (96, 97). Factors such as the degree of hydration and the size and concentration of counterions affect the secondary structure of the helix and lead to transitions from one to another form. The primary sequence of the polynucleotide also affects its helical conformations; synthetic duplexes having highly repetitive sequences possess quite distinctive helical forms (98). Metal complexes formed with these synthetic polymers may not reflect their native structures.

An important feature that distinguishes the two major classes of duplex structures and determines the helix conformation is the puckering of the furanose ring. A form DNAs and RNAs are characterized by C-3′ endo, or equivalent C-2′ exo, sugar puckering, while the B form exhibits C-2′ endo (or C-3′ exo) puckering of the sugar. Although this puckering may appear to be a small distinction between forms, it produces large differences in helical parameters, altering the helical pitch, the number of residues per turn, and the tilting of the base pairs away from the perpen-

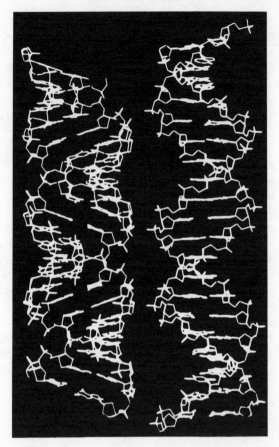

Figure 15   The A (left) and B (right) DNA duplexes.

dicular. These differences in turn are accompanied by changes in the width and depth of the helical grooves.

Computer graphics views of the A and B forms are shown in Figure 15, and parameters describing the various conformations are given in Table 6. The commonly observed B form of DNA is characterized by bases that are stacked perpendicular to the helix axis, located centrally with respect to the axis, and 3.4 Å part. With 10 residues per turn of helix, the pitch is 34 Å. The lack of base tilting and large pitch give rise to a wide (11.7 Å) major groove and a narrow (5.7 Å) minor groove of almost equal depths (8.5 and 7.5 Å, respectively). In the major groove, the positions N-7 and O-6(N-6) of the purine bases, and O-4(N-4) of the pyrimidines, are quite accessible to chemical attack.

Table 6 Helical Parameters of Various Polynucleotide Conformations[a]

Polymer	Form	Sugar Pucker	Pitch (Å)	Residues per Turn	Rise (Å) per Residue	Base Tilt (degrees)	Base Displacement (Å)[b]	Reference
DNA(Na)	A	C-3' endo	28.15	11	2.55	20.0	~5	95
DNA(Na)	B	C-2' endo	34.6	10	3.46	0	0	94
DNA(Li)	B	C-2' endo	33.7	10	3.37	2		93
DNA(Li)	C (B)	C-2' endo	31.0	9.3	3.32	6	1.5	89
Poly(dA)·poly(dT)	B'	C-2' endo	32.4	10	3.24–3.29	−7.9	> 3	98
Poly(dAT)·poly(dAT)	D (B)	C-2' endo	24.3	8	3.03	−8.0	−1.8	98
Poly(rA)·poly(rU)	A	C-3' endo	30.9	11	2.73	16.0	4	96
	A'	C-3' endo	36.0	12	3.02	10.0	4	96
RNA fragments	A"	C-3' endo	29–30	10–11	2.6–3.0			89
DNA–RNA hybrid(Na)	A	C-3' endo	28.8	11	2.62	20		89

[a] Adapted from reference 89.
[b] Displacement of center of base pair from helix axis.

The A class of DNA conformations is grossly different in structure. The pitch of 28 Å is 18% smaller, and there are 11 residues per turn, giving rise to a wider and more loosely assembled double helix. In the A double helix, as a result of the sugar puckering, the bases are tilted by as much as 20° from the plane perpendicular to the helix axis. In addition, the center of the base pairs is constrained to move approximately 5 Å away from the helix axis. The large base pair tilt and displacement leads to a very shallow (3 to 4 Å) minor groove and a deep (12 to 13 Å) major groove. The width of the major groove depends on the pitch of the helix, while the minor groove maintains a constant width of approximately 10 Å. As the pitch decreases, the bases tilt away and the major groove becomes quite narrow (10 to 20 Å).

RNA double helices assume only forms like that of form A DNA. The presence of the 2'-hydroxyl group requires a C-3' endo sugar pucker since in the C-2' endo position the 2'-hydroxyl group would point downward toward the bases and sterically hinder base-pair stacking interactions. Not surprisingly, DNA-RNA hybrids also exhibit only A-type conformations. Triple-stranded RNA helices have also been characterized (99), although their physiological role is uncertain. The triple helix of poly (U)·poly(A)·poly(U) is formed from the double helix by the addition of a poly(U) chain positioned parallel to poly(A) along the major groove of the helix and base paired through the N-7 and N-6 positions of the adenine ring.

3.1.2 Transfer RNA. The structure of tRNA, shown in Figure 16, affords the most complete description of a nucleic acid polymer. X-Ray diffraction analysis (100, 101) of single crystals of this small polynucleotide, as compared to the fiber diffraction studies of DNA and RNA duplexes, elucidated its structure in the solid state. This L-shaped polymer has both helical and looped regions, exemplifying the diversity of secondary and tertiary structure a nucleic acid can possess. A variety of non-Watson–Crick base-pairing schemes and unusual sugar puckering conformations are observed. The structure of tRNA and its binding by metal ions are examined in detail in Chapter 4.

3.1.3 Superhelical DNA. DNA isolated from a variety of sources exists not as a linear duplex chain but as a closed circle. Closed circular DNA is notable because of its great compactness in solution. A helical duplex joined end to end into a circle is subject to new topologic constraints. The compactness of the closed circular DNA has been attributed (102) to supercoiling brought about by these constraints. An illustration of superhelical DNA is given in Figure 17. In a closed circle the total number

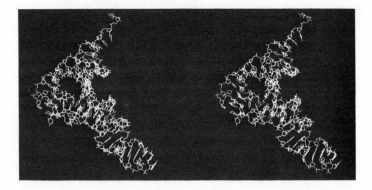

Figure 16 Stereoview of yeast phenylalanine tRNA. (Reproduced with permission from ref. 100.)

of helical turns (α), consisting of the number of Watson–Crick duplex turns (β) and superhelical tertiary turns (τ), must be a constant ($\alpha = \beta + \tau$). When the sugar phosphate backbone of even one strand is broken (or nicked), the constraints are relaxed, the superhelix unwinds, and the DNA forms an open circle having only duplex turns. In the cell, relaxing proteins (103) alter the topological winding number (α) of the DNA. These proteins nick the supercoiled DNA, allowing the reduction in superhelicity by an integral number of turns, and then reclose the circle. The result is a supercoiled structure with fewer superhelical turns.

Given the sensitivity with which small changes in tertiary structure can be monitored (104), these supercoiled DNAs are well suited for binding studies with small molecules. Even a reduction by one superhelical turn, causing a small decrease in the compactness of the structure, can be

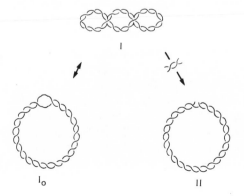

Figure 17 Illustration of closed circular DNA having several superhelical turns (I), relaxed closed circles having no superhelical turns (I_o), and a nicked circular DNA (II). [Reproduced with permission from *Acc. Chem. Res.*, **11**, 211 (1978). Copyright by the American Chemical Society.]

accurately assessed by gel electrophoresis (105, 106). Backbone cleavage, brought about by the binding of a metal complex, may be easily assayed since one nick in the duplex completely opens the supercoiled structure.

3.1.4 Chromatin. Within the nuclei of higher organisms, DNA is packaged into nuclei acid protein particles called chromatin (107). In partially denatured form the structure resembles "beads on a string," where repeating units of nucleosome particles (the beads) are linked by variable-length strands of DNA. The nucleosome cores contain a defined length of DNA (approximately 146 base pairs) and an octamer aggregate of basic histone proteins. Thus the anionic character of the polynucleotide is substantially neutralized in the presence of the protonated amine substituents of the histone proteins.

The structure of the nucleosome core is currently a question of active research. Small crystals have been obtained and analyzed (108) by electron microscopy and X-ray diffraction. The results indicate that the nucleosome core has a flat, wedge-shaped structure of dimensions 110 × 57 Å, with DNA surrounding the histone center. In the model proposed, the DNA is wound about the histones in 1.75 turns of a flat superhelix having a pitch of 28 Å.

The structure and reactivity of these protein nucleic acid particles should be taken into account in considerations of the in vivo interactions of heavy metal complexes with nucleic acids. The actual structure and variations in structure of the DNA throughout the life cycle of the cell remains unclear, however.

3.2 Nonspecific Metal Binding

The binding of metals to the polynucleotides may, in a manner similar to that observed in the simple associations of metal ions with the nucleotide monomers, be classified as occurring predominantly on the base or the phosphate or through stacking interactions. Metal binding to the polymer will also have a dramatic effect on the overall stability of the duplex structure. Binding to the phosphate groups, where charge neutralization decreases the electrostatic repulsions between neighboring nucleotide units, and metal complex intercalation, where base–ligand stacking interactions lower the overall free energy, both stabilize the double helix. In contrast, metal binding to the bases will usually disrupt base pair hydrogen bonding and destabilize the helix. While much research emphasis has been placed on finding selective metal reagents as probes of polymer structure, these nonspecific metal ion interactions demonstrate the variations in polymer structure and properties one is likely to encounter in the associations of metal complexes with polynuceotides and serve as a guide for the design of specific probe reagents.

3.2.1 Binding to the Phosphates and Bases. The stability of the double helix is largely a function of the type and concentration of counterions present. Repulsive forces between neighboring phosphate anions tend to unwind the helix in the absence of stabilizing counterions. Mono- and divalent cations that bind to the phosphate groups neutralize this electrostatic repulsion and stabilize the helix. The expansion or compression of the duplex is a function of cation concentration and has been detected by a number of techniques (109, 110). Relative helix stability can also be determined by examining changes in the helix-coil transition temperature, T_m, of the double-stranded polynucleotide (111). As the DNA melts and the two strands coil randomly, the bases unstack, leading to a cooperative increase in the absorption intensity at 260 nm. This hypochromicity, monitored as a function of temperature, is a measure of helix stability. It has been shown (112) that the T_m varies logarithmically with ionic strength; the higher the concentration of phosphate binding counterions, the greater the energy required to denature the helix and the higher the T_m.

This parameter is also a useful indicator of the mode of association of metal cations with the polyanion. In binding to the bases, metal ions often hinder the intrabase-pair hydrogen bonding interactions, lowering the T_m of polynucleotide. Figure 18 displays the T_m of a calf thymus DNA solution in the presence of a series of metal ions at various metal:DNA-phosphate ratios (113–115). These results parallel the relative binding patterns of the transition metals to the nucleotide monomers. Those metals that associate highly with the phosphate moiety of the nucleic acid produce a corresponding increase in the T_m of calf thymus DNA. Those metals that bind predominantly to the bases decrease the T_m, destabilizing the helix. It is noteworthy that in all cases shown in the figure there is an initial increase in the T_m at low metal:DNA-phosphate ratios. As has been stressed previously, the primary interaction of metals with the polyanionic nucleic acid is electrostatic in nature. At low metal:DNA-phosphate ratios, the transition metal ion binds exclusively to the phosphate anions, resulting in an increased melting temperature. At higher ratios, the binding of the metal to the bases can become more significant, destabilizing the helix relative to the coiled structure and lowering the T_m. The melting temperature is, then, seen to reflect the relative binding affinities of this series of divalent metal ions for the phosphate and base moieties of the polynucleotide. The preference for phosphate over base association decreases in the order Mg(II) > Co(II) > Ni(II) > Mn(II) > Zn(II) > Cd(II) > Cu(II).

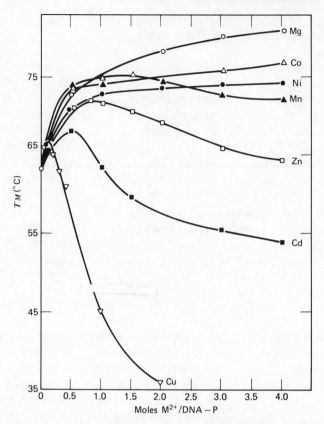

Figure 18 Effect of various metal ions on the melting temperature (T_m) of calf thymus DNA. [Reproduced with permission from *J. Am. Chem. Soc.*, **90**, 7323 (1968). Copyright by the American Chemical Society.]

This series could be extended to include the softer heavy metal ions as well. In the presence of either Pb(II) (116) or Au(III) (117), the melting temperature of DNA is substantially lowered. Heavy metal ions may therefore be placed on the far right of the series shown above. Actually the transition temperature does not directly distinguish base versus phosphate binding but indicates instead whether or not the helix is stabilized. Metal ions which, in binding to the bases, crosslink and stabilize the helical structure will cause an increase in T_m. An example of this crosslinking phenomenon is seen in the interaction of silver ion with DNA (118, 119). Potentiometric and spectrophotometric titrations indicate a strong pH dependent binding of Ag(I) to DNA. Also, the buoyant density of the DNA in a cesium sulfate gradient increases linearly with bound

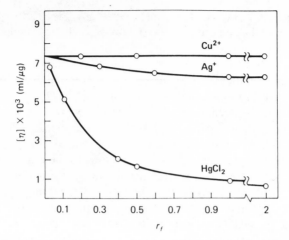

Figure 19 Effect of various metal ions on the intrinsic viscosity [η] of calf thymus DNA in 0.1 M NaClO$_4$ at 25.0°C, where r_f is the ratio of added metal/DNA-phosphate. [Reproduced with permission from *Biochim. Biophys. Acta,* **55,** 609 (1962). Copyright by Elsevier/North-Holland Biomedical Press.]

silver ion. The cooperative association of Ag(I) with duplex DNA and the observation that denatured DNA binds silver more strongly than the native form indicate the high binding affinity of Ag(I) for the base moiety. Based on proton release measurements, which show the dissociation of one proton per silver ion bound, a model for the structure of the silver–DNA complex was proposed involving the conversion of an N–H · · · N hydrogen bond of a complimentary base pair to an N–Ag–N bridge. A guanine–cytosine base pair would be linked both through an O-6 · · · N-4 hydrogen bond and a silver ion bound to the deprotonated N-1 of guanine and the N-3 of cytosine. The proposed complex bears a striking resemblance to the subsequently isolated silver cytosine dimeric complex discussed above (Fig. 11). Although the increased bond distances and helix diameter of the structure would presumably increase the flexibility of the duplex, silver ion, in linking the bases, would tend to stabilize the helical structure.

Another method of assaying the effects of metal ions on the duplex structure is by examining the hydrodynamic properties of the polynucleotide. Figure 19 displays the intrinsic viscosity of a calf thymus DNA solution in the presence of Cu(II), Ag(I), and Hg(II) as a function of the metal:DNA-phosphate ratio (118). Both the cupric and silver ions produce small reductions in solution viscosity, indicating a decrease in the overall stiffness of the duplex. In the presence of mercuric ion (120, 121), the intrinsic viscosity of the DNA solution is dramatically lowered. Thus

Hg(II) seems to alter grossly the duplex structure; the viscosity is roughly that of a completely denatured polynucleotide coil. It is unclear however that the structure of the heavily mercurated polynucleotide is fully collapsed. The hyperchromicity of the reaction product, compared to that of denatured DNA, indicates significant though limited base stacking. In kinetic and equilibrium binding experiments (122) of Hg(II) with poly(A)·poly(U), two reaction phases are observed. A fast second-order binding showing no cooperativity, but with the release of protons, is followed by a slow, cooperative, first-order process. A model to explain this result was proposed based on comparisons to the reaction rates with the single-stranded homopolymers. Initially the mercuric ion may bind the N-3 position of the uracil base with possible crosslinking to the opposite adenine. This intermediate structure would stabilize the duplex. Given the greater affinity of Hg(II) for uracil rather than adenine, a second bond to another uracil base then slowly forms, disrupting the duplex entirely and producing single-stranded loops with uracil bases bridged by the mercuric ion.

Soft heavy metal ions bind predominantly to the softer sites on the polynucleotide bases. The high mutagenicity of heavy metals (see Chapter 3) is most likely a function of this strong affinity for the base moiety. The results of binding experiments of heavy metals with polynucleotides are consistent with the site preferences observed in studies of heavy metal interactions with the mononucleotides. A new factor emerges in polymer interactions, however. In binding to the exocyclic or ring substituents of the bases, base-pair hydrogen bonding may be disrupted, with the bases being forced to unstack to accommodate metal ion binding. As the duplex opens, new favorable sites become available for metallation and the strands uncoil further. Alternatively the heavy metal may then covalently link base pairs, fixing the metallated helix into a stable form.

The binding of organomercurials to *E. coli* tRNAs containing thiobases illustrates another feature of metal interactions with polynucleotides. Given the strong preference of heavy metals for a "soft" sulfur atom, *p*-chloromercuribenzoate (PCMB) was expected to bind to the single 4-thio-U base of *E. coli* tRNA^{Val} at low metal–base ratios. No binding of PCMB to this tRNA was observed (123) however under native conditions. In the native structure, this "soft" site on the tRNA is apparently inaccessible to the bulky organomercurial. Equilibrium dialysis experiments (123) revealed that if the tRNA was partially denatured by the removal of Mg^{2+} and heating to 40°C, PCMB binds the 4-thio-U base with a high association constant and at the expected 1:1 stoichiometry. Furthermore, when Mg^{2+} was added back to the solution, spectral evidence showed that the tRNA returns to its native form and the metal reagent is released.

Similar results were found with other sulfur-containing tRNAs (124). These results demonstrate the importance of the nucleic acid tertiary structure in determining the coordinating ability of a base. Moreover, they reveal that the energy gained in stabilizing the native structure of a nucleic acid may prohibit reactions considered likely based on metal interactions with mononucleotides.

In summary, nonspecific metal ion binding to the polymeric duplex can be considered as a composite of interactions with mononucleotides. There is often an additional large alteration in the secondary and tertiary structure of the nucleic acid polymer. Finally, in some cases it appears that maintenance of the polynucleotide conformation is sufficient to restrict the otherwise favorable coordination of metal ions. These factors must be taken into account in the examination of heavy metal interactions with the polymeric nucleic acid.

3.2.2 Metallointercalation Reagents. Aromatic dyes such as proflavine, acridine, and ethidium bromide bind to double-stranded DNA in an intercalative (125) fashion, where the planar heterocycle is sandwiched between the base pairs of the helical polymer. This nonspecific stacking interaction stabilizes and stiffens the double helix. The polymeric chain is lengthened to accommodate the cationic dye. Metal complexes can also bind primarily through intercalation. We have seen how stacking forces can stabilize the association to mononucleotides with metal complexes. Similarly, intercalative binding accounts for the large affinity for double stranded polynucleotides of planar, cationic metal complexes having heterocyclic ligands. Indeed, the metal center may enhance stacking interactions. Platinum(II) complexes have been long realized to stack in columnar arrays in the solid state and are especially good intercalators.

The intercalative binding of (terpyridine)platinum(II) complexes to nucleic acids, proposed initially (126, 127) to account for results of binding experiments of [(terpy)PtCl]$^+$ with tRNA, has been demonstrated (127, 128) through a variety of techniques. In the presence of [(terpy)Pt(HET)]$^+$, 2,2′,2″-terpyridine-2-hydroxyethanethiolatoplatinum(II), there is an increase in the melting temperature of calf thymus DNA, reflecting the increased stabilization of the helix versus the coiled form of the duplex because of the additional intercalative stacking interactions. Figure 20 displays a plot of the viscosity of calf thymus DNA as a function of increasing concentrations of bound [(terpy)Pt(HET)]$^+$. Unlike the heavy metal ions that bind to the base substituents and denature the DNA, decreasing its viscosity, the platinum reagent induces an increase in the viscosity of the DNA solution. As the polynucleotide binds the platinum complex, the helix unwinds, lengthens, and stiffens until the

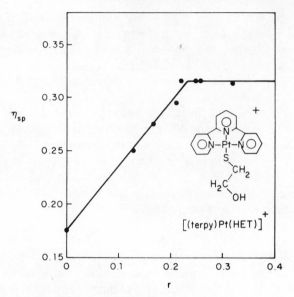

Figure 20 Specific viscosity (127) of calf thymus DNA at 25°C as function of increasing amounts of bound [(terpy)Pt(HET)]$^+$ per phosphate, r.

saturation limit of binding at a ratio of ~0.22 platinum cations per DNA-phosphate group.

Intercalative binding may also be demonstrated through competitive binding experiments using ethidium bromide, a well-studied aromatic dye which shows enhanced fluorescence on intercalation. The fluorescence of an ethidium–DNA solution will be diminished in the presence of a competing platinum intercalating reagent. The binding may be approximately fit to the Scatchard equation (129), $r/c = K(n - r)$, where r is the ratio of bound ethidium/DNA-phosphate, c is the concentration of free ethidium, n is the number of available binding sites, and K is the binding constant. In the presence of a competing reagent that can occupy the same site on the duplex as the ethidium cation, the effective binding constant of ethidium to DNA, measured through the fluorescent intensity and given by the slope of a Scatchard plot, is reduced. If the test reagent binds covalently to the DNA, altering the nature of the intercalation site, the number of remaining available sites for ethidium, n, given by the ordinate intercept of a Scatchard plot, is lowered. Figure 21 shows the results of such a competition experiment using [(terpy)PtCl]$^+$. At low concentrations the platinum reagent competitively inhibits ethidium intercalation. At higher concentrations, where the substitutionally labile chloride ion may dissociate, the [(terpy)PtCl]$^+$ complex covalently binds to the bases, non-

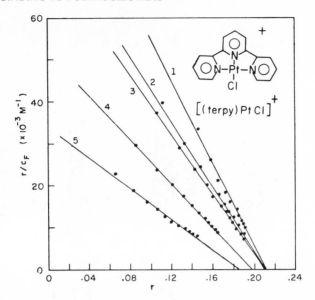

Figure 21   Fluorescence Scatchard plots of ethidium binding (1) to calf thymus DNA and in the presence of increasing concentrations (2–5) of [(terpy)PtCl]⁺. See reference 128 for more details.

competitively inhibiting ethidium intercalation. This assay has been used to investigate (128) both the covalent and intercalative binding modes of a series of planar platinum complexes. Platinum cations having heterocyclic ligands such as *o*-phenanthroline, 2,2′-bipyridine, and 2,2′,2″-terpyridine bind noncovalently to DNA through intercalation. In addition, hydrogen bonding interactions between coordinated ligands and the phosphate backbone stabilize the stacked complex. The structure of the intercalated [(terpy)Pt(HET)]⁺ −dCpG complex, for example, shows the hydroxyethanethiolato tail to be hydrogen bonded to the phosphate oxygen atom (Fig. 13).

Closed circular DNAs may also be employed (127, 130) to demonstrate intercalative binding of platinum reagents. Nicked and relaxed closed circular DNAs having no superhelical turns will comigrate in assays for their distinctive shape, such as gel electrophoresis or sedimentation velocity experiments. In the presence of the platinum intercalating reagent, however, the relaxed closed DNA will wind into a superhelix. The duplex must unwind to accommodate platinum binding, and, given the topological constraint, a reduction in duplex turns must coincide with an increase in superhelical turns. These same constraints do not apply to the nicked form. Hence the binding of the intercalating reagent results in their sep-

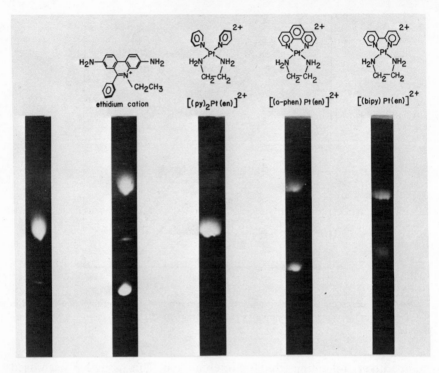

Figure 22 Electrophoresis in 1% agarose gels of relaxed closed circular and nicked circular PM-2 DNAs (130), where (from left to right) the gels contained no added reagent or contained ethidium bromide, $[(py)_2Pt(en)](ClO_4)_2$, $[(o\text{-phen})Pt(en)](NO_3)_2$, or $[(bipy)Pt(en)](NO_3)_2$. After electrophoresis, all gels were stained with ethidium bromide. As a result of intercalative binding, the nicked circles have a reduced electrophoretic mobility, while the closed circular DNA, having an increased superhelicity on binding the intercalator, migrates more rapidly. The relaxed closed and nicked circular DNAs comigrate in the absence of added intercalating reagents.

aration under conditions that assay for superhelicity. Figure 22 displays the gel electrophoretic pattern of nicked and relaxed closed DNAs in the presence of a series of platinum reagents having different ligands. In the presence of $[(o\text{-phen})Pt(en)]^{2+}$, $[(bipy)Pt(en)]^{2+}$, and ethidium, the closed circle migrates farther, reflecting its compact, supercoiled structure. In the presence of $[(py)_2Pt(en)]^{2+}$, by contrast, the nicked and closed circles comigrate. This last platinum reagent cannot intercalate into the DNA. Nonbonded steric interactions force the pyridine rings out of the platinum coordination plane, thereby precluding intercalation. The study of these metallointercalation reagents and the nonintercalating analog therefore provides a nice demonstration of the requirement of planarity for inter-calative binding.

The tendency of planar platinum(II) complexes to form one-dimen-sional columnar stacks in the solid state, the aromaticity and size of the heterocyclic ligand, the positive charge, and the planarity of the complex all contribute to the ability of these platinum complexes to bind intercalatively to nucleic acids. The substitutionally inert [(terpy)-Pt(HET)]$^+$ complex intercalates both into DNA and RNA helices. The unique cooperative association (131) of [(terpy)Pt(HET)]$^+$ with RNA duplexes, as opposed to its anticooperative binding to DNA, reflects the differences in the secondary structures of these polynucleotides. Fur-thermore, platinum metallointercalation reagents show (132) a preference for binding to DNAs having a high guanine–cytosine content, indicating a degree of specificity in the bound intercalation site. As we see below (Section 3.3.2), these heavy metal reagents have proved quite useful in elucidating the structure of the DNA helix saturated with bound inter-calator.

3.3 Heavy Metal Probes of Polynucleotide Structure

An understanding of the nonspecific associations of heavy metals with polynucleotides and the selective reactions of these metals with the monomers provides a basis for the design of specific heavy metal reagents to probe nucleic acid structure. The high electron density and relative inertness of heavy metal complexes allow for their use as markers in a variety of physical assays. Heavy metals, because of their high electron density, can be used as specific labels both in electron microscopic and X-ray diffraction experiments. The high stability, specificity, and in-creased buoyant density of complexes formed with these reagents permit their utilization in the purification and separation of nucleic acids. Also, the fluorescence of lanthanide ions provides a new tool for the elucidation of polymer structure.

3.3.1 Reagents for Electron Microscopy.

Heavy metals such as platinum, palladium, and uranium have long been used as nonspecific polymer stains in electron-microscopic work (133). The scattering power of these electron-dense reagents substantially improves the contrast of images in the electron microscope. A variety of metal clusters and oligomers have been employed in low-resolution work. The protein ferri-tin, containing approximately 2000 iron atoms in its core, provides a large electron-dense label for investigations (134) of cellular structure. Re-cently, the platinum pyrimidine blues have also been shown (135) to be excellent polynucleotide stains for studying cytological fine structure. Indeed, rapid advances in the fields of molecular microscopy and heavy metal chemistry now permit the labeling of specific sites on a polynucleotide in aqueous solution for electron-microscopic analysis of structure.

A goal toward which much research in this field has been directed is the sequencing of DNA and RNA in the electron microscope through the use of heavy metal labels to distinguish each of the bases (136). After specifically marking a particular set of nucleotides, for example all the adenines, their positions on the polymer could in theory be read off an electron micrograph. Given the availability of quantitative, nucleotide-specific reagents, the primary sequence of the nucleic acid could be easily determined in two labeling experiments. What this method requires however is exceedingly high-resolution microscopy. In order to distinguish neighboring base markers, ~5 to 7 Å resolution of single heavy atoms or, possibly, of polynuclear clusters is necessary. While both scanning transmission (137) and dark-field (138) electron microscopy have advanced to the stage where single heavy atoms can be detected, from a practical viewpoint the current status of electron microscope technology remains a major obstacle to the method.

The other important prerequisite for this technique is the availability of highly selective, stable, heavy atom labels of the nucleotides. Labeling schemes, to prove applicable, have to meet a number of chemical criteria. Firstly, reaction conditions must be sufficiently mild to maintain the integrity and solubility of the polymer. Conditions which result in the degradation, depurination, or precipitation of the polymer are prohibitive. The moderate water solubility of heavy metal reagents has proved advantageous in this regard. Secondly, quantitative binding of the reagent(s) to each particular nucleotide without side reactions with the phosphate or base moieties of other nucleotide units must be achieved. An uncertainty of 0.1% in the labeling of any one base would result in a confidence of only 90% in the fidelity of labeling a 100-nucleotide fragment. Added to this uncertainty is the fact that, in the electron microscope, one is examining individual molecules rather than an average species distribution. The complementarity of the DNA duplex does however provide a valuable control. The sequencing method also requires that the electron-dense marker be at a defined distance with respect to its associated nucleotide so as to reflect truly the base position. Approaches to reagent design have included the synthesis of extended chain ligands with a cluster of heavy atoms at the tail. Since neighboring bases are only a few angstroms apart in the duplex, however, an accurate reading of the micrographs necessitates unambiguous heavy atom labeling at close proximity.

One method to achieve base specificity is through reagents that distinctly bind each of the different base moieties. The addition of mercuric acetate to the nucleotides results (56) in the selective metallation of UTP and CTP at the C-5 position. Quantitative acetoxymercuration of poly(U) has also been demonstrated (139). Mercurated poly(C) precipitates after 20 to 40% reaction, while no binding of mercury is found to either poly (A), poly (G), or poly (T).

Platinum complexes, because their primary sites of interaction are with the bases and because they are relatively inert to substitution, provide useful labels for the polynucleotides. The binding specificities of K[PtCl$_3$(DMSO)] with both homopolymers and native DNA have been examined (140). This anionic reagent should bind monofunctionally to the available nitrogen atoms of the bases, given the trans labilizing influence of the sulfur-bonded DMSO ligand. The product formed, PtCl$_2$(DMSO)L, would be neutral and would therefore not be expected to alter the solubility properties of the polymer. Neither crosslinking, which could result from bifunctional association of the reagent with two different bases, nor precipitation (at 0°C) appeared as significant problems in these studies, although crosslinking of the duplex was not specifically assayed. The reactions were found to be kinetically slow and pH dependent, but native polynucleotides showed an enhanced rate of reactivity. The following levels of platinum binding were observed: 2 platinums/adenine; 1 platinum/cytosine; 0.6 to 0.8 platinums/uracil, and 2 platinums/guanine at pH 6.0 (1 platinum/guanine at pH 7.5). Binding to uracil was dependent on the polymer conformation. Presumably metalation occurs at the N-7 and/or N-1 nitrogen atoms of the purines and the N-3 nitrogen atom of pyrimidines. While not actually a selective marker for particular bases, this reagent may prove to be a useful stain for the adenine and guanine moieties of a polynucleotide. Since the binding of this reagent denatures the polymer, it can be used only as a stain for single-stranded species, the conformations and base orientations of which are certain to be less regular than those of the duplex.

Another class of extensively investigated heavy metal reagents are the pyridine and bipyridine derivatives of osmium tetroxide (141). In the presence of added ligands such as cyanide or pyridine, osmium tetroxide reacts quickly and quantitatively both with carbon–carbon double bonds and *cis*-diol substituents (58), as we have mentioned previously. This reagent could therefore provide a kinetically fast, low-temperature, selective label for the pyrimidine bases of DNA through addition across the 5–6 double bond. Incorporation of an additional heavy atom into the pyridine ligand, for example, 3-acetoxymercuripyridine (142), enhances

Table 7   Quantitation of OsO₄-Bipyridine Binding to Polynucleotides[a]

Polymer	Reaction Time	Os/DNA-Phosphate Ratio
Poly (U)	24 h	0.97
	4 days	0.83, 0.91, 1.00
Poly (C)	24 h	0.86, 1.00
Poly (A)	24 h	0.0
Poly (G)	24 h	0.0
Denatured DNA	24 h	0.48, 0.48, 0.55, 0.49
	3 days	0.50
	13 days	0.49, 0.47, 0.49
Native DNA	21 h	0.45, 0.44

[a]Reproduced from *Biochemistry*, **16**, 36 (1977). Copyright by the American Chemical Society.

the electron density of the reagent. One problem encountered with this label however was that the bis(pyridine)osmate esters undergo transesterification reactions (58, 143). The resulting ligand exchange in the polynucleotide binding experiments removed the pyrimidine labels. More recently, bipyridine has been shown (141) to stabilize the osmate ester products and the use of OsO₄/bipy has resulted in 90% labeling of all pyrimidines in calf thymus DNA. Table 7 displays the results of binding experiments with synthetic homopolymers as well as native and denatured DNA. Negatively charged nitrogen chelate ligands, such as 4,4'-dicarboxy-2,2'-bipyridine, have also been explored. The bulky ring substituents hinder intercalative stacking interactions that might otherwise complicate the interpretation of electron-microscopic results. Furthermore, the anionic labels tend to repel each other, leading possibly to larger interbase separations on the grid. Scanning transmission electron microscopy (STEM) results, where poly(U) and poly(dAT) were labeled with these reagents, have been reported (144).

One method attempting to extend the utility of osmium tetroxide as a probe reagent involved chemical modification (145) of the bases to include additional olefinic sites for metalation. Chloroacetaldehyde reacts with cytosine and adenine nucleosides to yield the etheno derivatives, 1,N³-ethenocytosine and 1,N⁶-ethenoadenine. Calf thymus DNA derivatized in this manner binds one osmium atom per nucleotide phosphate, consistent with the addition of one osmium atom per 1,N⁶-ethenoadenine and -thymine and two per 1,N³-ethenocytosine. An alternative modification scheme for cytosine involves the formation of a furyl derivative through reaction with O-furfurylhydroxylamine and bisulfite. The furan ring is susceptible to the binding of two osmium or three mercury atoms. Simi-

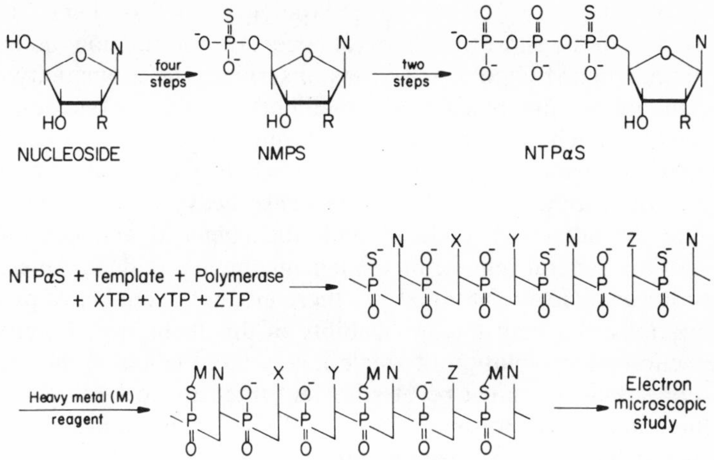

Figure 23 General scheme for the specific heavy metal labeling of polynucleotides. [Reproduced with permission from *Acc. Chem. Res.*, **11**, 211 (1978). Copyright by the American Chemical Society.]

larly, 2-furylglyoxal has been shown to react quantitatively with guanine nucleotides.

While these base-specific reagents and base modification schemes provide a viable route for the selective labeling of the nucleotides, an alternative approach for differentiation based on a modification of the common phosphate moiety of the nucleotides was developed following the discovery (146) that phosphorothioate groups can be enzymatically incorporated into RNA and DNA. Given the high affinity of heavy metal atoms for sulfur, the labeling scheme depicted in Figure 23 was devised (147). After transcription of the polynucleotide under study, using one specifically thiolated and three normal nucleoside triphosphates (for example, ATPαS, GTP, CTP, and UTP) as substrates, monofunctional heavy metal reagents could be attached selectively to the phosphorothioate group on the polymer backbone adjacent to the specified base (adenine in this case). The advantages of this technique lie, firstly, in the fact that the electron-dense marker would be situated directly on the polymer backbone and, secondly, that each of the bases could be labeled using the identical experimental design. Moreover, highly specific and stable binding of a variety of heavy metal reagents to the soft sulfur atom of the phosphorothioate group can be achieved. All that is needed for this modification scheme is a source of phosphorothioated mononucleotides. The uncertainty in fidelity introduced in the additional transcriptional step is known to be low. In a double-label experiment, the quantitative binding of $[^3H][(terpy)PtCl]^+$ to $[^{35}S[poly(_sA-U)$ has been demonstrated (147). In the short time interval necessary for reaction, no binding to the unmodified $[^{14}C]poly(A-U)$ was observed. The aromaticity of the terpyridine ligand may possibly enhance both the rate of reaction of the heavy metal reagent and the stability, through intercalative stacking, of the product formed. The phosphorothioate labeling technique is currently being studied with a series of electron dense heavy metal reagents.

Despite the advances made in both the chemical and instrumental aspects of the general scheme of sequencing nucleic acids in the electron microscope using heavy atom labels, there are still a number of practical considerations that limit the applicability of this technique. Foremost is the detection and resolution of single heavy atom labels in the electron microscope. Markers must be of sufficient intensity to be distinguished from the background noise of the grid. While background filtering and averaging techniques have been applied to this problem (144), better grid preparations and labels of higher electron density are desirable. A number of laboratories have synthesized large, water-soluble, polynuclear clusters for use in the electron microscope. The binding of these reagents to polynucleotides has however infrequently been examined. Resolution in the electron microscope further limits possible labeling schemes that depend on the number of heavy atoms for differentiation. Distinguishing an adenine base marked with three heavy atoms from a cytosine base marked with two heavy atoms is not feasible at the present time. In addition, it is not clear that the marked positions observed in the electron micrograph are at all representative of the polynucleotide structure and hence reflective of the polymer sequence. The high-energy electron beams used (approximately 20 kcal/$Å^2$) severely damage the specimen and cause the motion of the heavy atoms on the grid. Low-temperature microscopy aids but does not alleviate this problem. Moreover, recent evidence (148) suggests that the structure of the native B–DNA duplex, characterized by a 3.4 Å rise per residue, differs from the structure observed in the electron microscope (2.9 Å). While these results are not surprising, they do not bode well for a labeling scheme using the electron microscope which requires an accurate assessment of base position. Finally, the applicability of this sequencing method is limited by the availability of high-resolution (STEM) electron microscopes; very few currently exist. Rapid progress has been made in the sequencing of nucleic acids through chemical and enzymatic methods (149, 150) using gel electrophoresis. Given the practical utility and wide acceptance of this method, the goal of sequencing DNA and RNA in the electron microscope should perhaps be reappraised.

The value of heavy-atom labeling of nucleic acids rests in its use as a tool to obtain biological information by specifically marking in solution sites on the polynucleotide for electron-microscopic analysis. Various applications of the above and other reagents to problems other than sequencing are currently being explored. For example, ribosomal binding sites on 16S RNA have been examined (151) through heavy-metal polynucleotide labeling. In addition, the electron-dense mercury atom cluster tetrakis(acetoxymercuri)methane has been shown (152) to bind specifically to the unique 4-thiouridine residue of *E. coli* tRNA^{Val}. Polymercurated reagents of this kind should be visible in conventional electron microscopes requiring lower beam intensities. As a consequence of the research in this area, heavy metal reagents can now provide investigators with general or specific electron-microscopic probes of biological structure. The future appears even more promising.

3.3.2 Reagents for X-Ray Diffraction. Electron-dense heavy metal atoms are important tools in X-ray crystallographic determinations of biopolymer structure. In order to solve the structure, isomorphous heavy metal derivatives of the crystalline polymer are often employed for the assignment of phases. Since tRNA thus far serves as the only example of a crystalline polynucleotide, the number of examples where heavy metal reagents have served as isomorphous replacements in nucleic acid structural studies is accordingly limited. A variety of structural illustrations of heavy metal interactions with a nucleic acid polymer are nonetheless provided by the tRNA work. Heavy metal complexes have also proved useful in fiber diffraction studies characterizing the intercalative mode of binding. As is seen below, the distribution of bound drug along the polymer can be probed using an electron-dense metallointercalating reagent.

Using the refined atomic coordinates from the crystal structure of yeast tRNA^{Phe}, the strong binding sites of a series of metal reagents have been identified (153). The heavy metal derivatives, for the most part, were prepared by soaking the native crystalline tRNA in a solution containing the particular metal complex. The binding modes are consistent with expectations based on the knowledge of small molecule crystal structures. The soft, heavier metals bind predominantly to the heterocyclic nitrogen atoms of the base moieties, while the lighter transition metal ions coordinate both to phosphate oxygen atoms and to sites on the bases. Magnesium and samarium—hard, electropositive ions—bind extensively to the phosphate groups, crosslinking and stabilizing the tertiary loops of the polymer backbone.

The square planar *trans*-dichlorodiammineplatinum(II) reagent binds strongly to tRNA at a specific site, producing an excellent heavy metal derivative. In contrast, the cis isomer, highly studied as an antineoplastic agent, is not a good isomorphous replacement reagent. Electron density maps show the *trans*-diammineplatinum(II) moiety coordinated to the N-7 nitrogen atom of the guanine residue Gm34 located on the anticodon loop (See Chapter 4). One ammine group is oriented so as to hydrogen bond to the O-6 oxygen atom of the same residue, and the trans ammine group forms hydrogen bonds to three oxygen atoms of the adjacent phosphate group. Perhaps these hydrogen bonding interactions, not available in the case of the cis isomer, explain the high specificity of the trans reagent for this binding site.

The osmium–bipyridine complex binds strongly to five distinct sites of tRNA. Three of the five sites, one of which is identical to the platinum site, involve direct metal coordination to the N-7 nitrogen atom of a guanine residue. These interactions are reinforced by hydrogen bonds between hydroxyl groups coordinated to the metal and acceptors on the base moieties. Another metalation site results from the formation of the cyclic osmate ester with the O-2′ and O-3′ oxygen atoms of a ribose group. The fifth labeled site may involve metal addition across the 5–6 double bond of a thymine residue, a reaction extensively utilized in electron-microscopic labeling schemes. That these varied binding modes are all present in osmium–tRNA complexes demonstrates the contribution of polymer conformation in determining the reaction product; the binding mode and position depend on the accessibility of sites in the extensively looped tRNA. An additional illustration is provided by the interaction of a mercury complex, hydroxymercurihydroquinone-*O,O*-diacetate, with tRNA. The site of mercuration is the O-4 oxygen atom of an exposed uracil residue, U47. This mode of attachment is similar to that observed in the monomeric complex between $HgCl_2$ and uracil (77).

Lanthanide derivatives, as might be expected, provide structural analogs of the interaction of magnesium ion with tRNA. Both samarium and lutetium cations bind strongly to phosphate oxygen atoms of the polymer backbone. In two of the metalation sites lanthanide ions crosslink pairs of phosphate groups. In bridging the phosphate backbone, the metal ion presumably stabilizes the tertiary structure of the tRNA. Indeed, in the native crystal structure where Mg^{2+} serves as counterion, these sites are occupied by tightly bound magnesium ions..

These crystalline heavy metal derivatives, used in the X-ray crystallographic determinations of the structure of tRNA, exemplify many of the specific modes of metal binding to nucleic acids and, additionally, the importance of polymer tertiary structure in determining the mode of

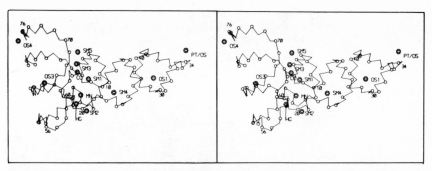

Figure 24 Stereoview of the tRNA backbone showing the binding sites for various metal reagents. [Reproduced with permission from *J. Mol. Biol.*, **111**, 315 (1977). Copyright by Academic Press, Inc. (London) Ltd.]

interaction. Figure 24 displays the various metal binding sites on the tRNA backbone. The hard phosphate-binding metals appear to occupy sites within the extensively looped core, bridging phosphate moieties, while the softer heavy metals tend to bind the exposed base residues near the periphery of the polymer. The only requirements to establish the suitability of a given heavy metal reagent as an isomorphous replacement are, firstly, that the metal ion does not disrupt the polynucleotide structure, thereby altering the crystal form, and, secondly, that the probe reagent binds strongly and specifically to a given site rather than randomly, which disorders the structure. Many of the electron-dense probe reagents discussed previously, including polynuclear clusters, should prove valuable in the solution of new polynucleotide and perhaps nucleic acid–protein crystal structures.

Stereochemical details of the DNA duplex structure in the presence of intercalating reagents have also been made evident through the examination of heavy metal derivatives in X-ray fiber diffraction experiments (130, 154). As shown in Figure 25, in the presence of a bound intercalator such as ethidium bromide (155), the regular "cross-hatched" fiber diffraction pattern, which is characteristic of the native B–DNA duplex, is disrupted. In order to accommodate the intercalator, the helix must unwind, locally disordering the helical pitch and the number of base residues per turn of helix. This disordering of the helical structure results in a decrease in the corresponding intensities of the diffraction pattern from that of the regular helix. Figure 25c displays the fiber diffraction diagram of calf thymus DNA with a bound metallointercalating reagent, $[(\text{terpy})\text{Pt}(\text{HET})]^{+}$. Here, too, the helical pattern is diminished, yet reflections arising from the ordered distribution of the electron-dense platinum intercalator are observed.

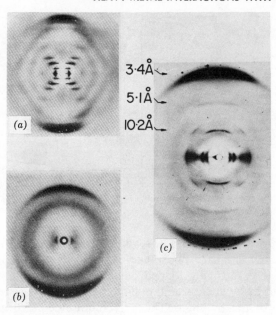

Figure 25 Fiber diffraction patterns of (a) sodium DNA in the B form at 92% relative humidity (155), (b) DNA in the presence of ethidium at 92% relative humidity (155), and (c) DNA in the presence of [(terpy)Pt(HET)]$^+$ at 98% relative humidity (154).

Scatchard plots, representing the binding of many intercalating drugs to the DNA duplex, reveal a saturation limit, n, approaching 0.25 molecules bound per nucleotide-phosphate. These plots have led to the proposal (156) that intercalative drugs bind the duplex in a "neighbor excluded" fashion, where every other interbase pair site contains bound intercalator at saturation. The diffraction pattern given in Figure 25c supports this neighbor exclusion model. As Figure 26 illustrates, occupation of sites on the duplex according to the neighbor exclusion principle results in repeated distances of 3.4 Å, the base pair stacking distance, and 10.2 Å, corresponding to the binding of one intercalator per two base pairs. Intercalating reagents such as ethidium (155) or proflavine (125) do not differ significantly in electron density from an average base pair; hence only a 3.4 Å periodicity in the diffraction pattern is observed. The presence of the electron-dense platinum atom in the intercalator, on the other hand, enhances the 10.2 Å periodicity. A 10.2 Å layer line, seen in the diffraction pattern given in the figure, is also observed (130) in patterns of calf thymus DNA in the presence of other bound intercalating reagents. These results lend strong support to the neighbor exclusion model of intercalative binding.

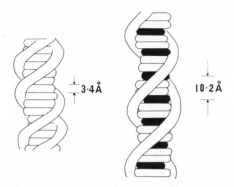

Figure 26 Schematic representation of a normal helix (left) and the neighbor-excluded binding (right) of an intercalator (shaded) to the duplex. [Reproduced with permission from ref. 154.]

The use of heavy metal probes in these diffraction studies afforded structure information about the duplex. Stereochemical models, describing the sugar–phosphate backbone conformation, were suggested based on these diffraction patterns. Fourier transform calculations indicated that the DNA is unwound 22 ± 6° to accommodate an intercalator. These diffraction experiments, then, represent another application where heavy metal reagents have been successfully employed to probe nucleic acid structure.

3.3.3 Reagents to Identify and Purify Polynucleotides. The chemical reactivity and physical properties of heavy metal complexes allow for their use in the isolation and characterization of nucleic acids. The increased buoyant densities of polymers with bound heavy metals, for example, can be put to advantage in purification schemes. The specificity with which heavy metals bind soft ligands makes them useful in chromatographic separations of polynucleotides. Moreover, mercurated nucleotides, which are substrates for RNA polymerase, have been used as labels in biological studies of RNA synthesis and processing.

As discussed previously (Section 3.2), the buoyant density of DNA, measured in a cesium sulfate density gradient centrifugation experiment, increases linearly with bound silver or mercuric ion. The level of binding at a given metal ion concentration depends upon the conformation of the polynucleotide, whether it is double or single stranded, and the base composition. Mercuric ion, for example, binds more strongly to AT-rich rather than GC-rich DNAs, while the reverse is true for silver ion. The separation of DNAs having different base compositions, or of native and denatured forms, can therefore be facilitated through density centrifugation experiments (157) based on these differences in binding specificities. Figure 27, for example, shows the densitometer tracings from an analytical ultracentrifuge experiment used to separate DNAs. In the absence of

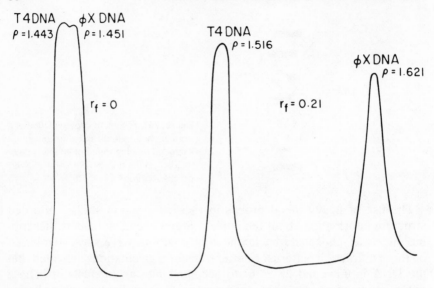

Figure 27 Densitometer tracings of an analytical ultracentrifugation experiment showing the separation of the single-stranded ϕX DNA from the double-stranded T4 DNA in the presence of Hg(II). [Reproduced with permission from *Biochemistry*, **4**, 1687 (1965). Copyright by the American Chemical Society.]

metal reagent, $r_f = 0$, the double-stranded T4 DNA and single-stranded ϕX-174 DNA sediment to approximately the same buoyant density, $\rho = 1.45$. At an added metal:DNA-phosphate ratio r_f of 0.21, however, the difference in buoyant densities between these double- and single-stranded species is enhanced, $\rho = 1.516$ and 1.621, respectively. The metal ion binds more strongly to the single-stranded ϕX-174 DNA, preferentially increasing the buoyant density of this polymer. It is important to note that since these metal ions bind reversibly, the polynucleotide may be isolated intact following separation. The addition of ligands having high affinity for the heavy metal, such as cyanide or mercaptide, dissociates the metal from the polymer. Native crab dAT, a minor component of crab DNA consisting primarily of the alternating copolymer poly(dAT), has been purified (158) by means of density centrifugation using mercuric ion to enhance buoyant density differences. DNAs rich in GC content have been shown (159) by density centrifugation to bind preferentially the platinum antitumor drug *cis*-DDP. The secondary structure of superhelical DNA has also been examined (160) in centrifugation experiments where methylmercuric hydroxide has been employed as a probe for unpaired bases. The heavy metal complex selectively binds to denatured regions of the duplex at sites where base pair hydrogen bonds are disrupted and favorable coordinating positions on the base residues are exposed. In this instance as well, the binding specificities of the heavy metal and the large buoyant shifts on binding to the polynucleotide lead to a sensitive assay of polymer structure.

The mercurated nucleotide 5-HgUTP has proved to be a valuable probe reagent (56, 139, 161) in both the purification and identification of RNA fragments. The complex can serve as a substrate for polymerization with a fidelity equal to that of the nonmercurated nucleotides (161). Moreover, the presence of the mercury substituent, positioned in the major groove of the helix, does not appear to alter significantly the structure of the duplex. These mercurated polynucleotides function efficiently as templates for polymerization. The stability of the mercury–carbon bond affords a relatively inert heavy metal label.

A primary advantage of these probe reagents is that in biological assays monitoring for example RNA synthesis, mercurated and nonmercurated polymers can be rapidly separated. One method of fractionation, based on the buoyant density increase, is discussed above. Alternatively, the separation of the metalated polymer can be achieved using chromatography resins containing free sulfhydryl groups. Rapid fractionation of the mercurated polynucleotides by this method has been demonstrated (161). The "sulfhydryl–agarose" affinity column selectively retains the metallated polymer, given the high affinity of heavy metals for sulfur. Recovery of the polynucleotide is achieved by elution with a competing mercaptan. The mercurated nucleotide 5-HgUTP is currently being employed as a substrate for RNA polymerase in in vitro studies (162–164) of transcription. Mercury substitution should permit the labeling and separation of newly synthesized transcripts.

3.3.4 A Fluorescent Probe.

Members of the trivalent lanthanide series luminesce in aqueous solution at room temperature. Since the lanthanide cations bind strongly to polynucleotides, their utility as spectral probes becomes evident. Both the binding of lanthanides and the competition of these cations with magnesium have been examined (165, 166). In binding to tRNA, the cation may serve as a useful fluorescent probe of polymer structure.

Free in solution, the lanthanide cations exhibit a low fluorescent intensity as a result of their small absorption coefficient. When bound near a highly absorbing species, however, the fluorescent intensity is substantially enhanced. Addition of the Eu(III) cation to *E. coli* tRNA under saturating conditions results (165) in a 300-fold enhancement in the europium fluorescence relative to that of the free ion. The concomitant decrease in emission intensity from the fluorescent 4-thiouridine residue

indicates energy transfer from that base to the lanthanide ion. Destruction
of the 4-thiouridine quenches the fluorescence of the bound lanthanide.
Similarly, the addition of the metal cation to a tRNA lacking the s⁴U
residue does not result in an increased fluorescent emission. A rapid
screening assay for the presence of this base in a given tRNA could
therefore be devised based on the presence or absence of s^4U sensitiza-
tion of lanthanide ion fluorescence.

There are at least four strong binding sites on the tRNA for Eu(III) or
Tb(III) cations. Competition experiments (165) suggest that magnesium
ion binds strongly at identical positions, consistent with the similar
affinities of magnesium and lanthanide ions for the binding of phosphate
groups. One metal binding site must be located in the vicinity of the s^4U
residue at position 8, since energy is transferred between these sites.
Indeed, in the X-ray diffraction study (153), another lanthanide cation,
samarium, has been observed to bind at this location.

While fluorescence is a property particular to lanthanide ions rather
than one shared by all heavy metal ions, its application to polynucleotide
structural problems remains to be fully realized. Fluorescent reagents
could be used to investigate heavy metal binding sites. Moreover, by
examining energy transfer between bound fluorescent labels, structural
information about nucleic acid particles could be obtained.

4   A PLATINUM ANTITUMOR DRUG

The discovery (1) by Rosenberg that *cis*-DDP exhibits antitumor activity
in a broad spectrum of cancerous tissues has greatly stimulated research
into the interactions of heavy metals with nucleic acids. The history
surrounding this discovery and the clinical utility of *cis*-DDP are described in the preceding chapter. What remains unclear is the mode of
action of *cis*-DDP. An understanding of the mechanism by which this drug
exerts its antitumor effects would surely aid in the design of new antineoplastic agents and perhaps provide hints as to the nature of cancer. In this
section we review some of the in vivo and in vitro experiments using this
drug and explore some structural models for the mode of action to which
these studies point. We begin by examining aspects of the biological
effects of the platinum agent and then focus in on features of the chemical
interaction that could determine the antitumor activity of this drug. Our
discussions thus far have laid the foundation for this examination. At the
outset one should be reminded of the difficulty in establishing a phar-
macological mechanism. The diversity of chemical targets within the cell as
well as the variety of reactions that we have shown heavy metal reagents

undergo lead to an abundance of possible metal–polymer structural products. The determination of which specific interaction is the "potent" one becomes quite difficult.

4.1 Cytotoxic Activity and Possible Modes of Action

4.1.1 Structure–Activity Relationships.   Some clues as to the mode of action of *cis*-DDP can be derived from structure–activity studies (167–169) of various platinum analogs. Table 8 summarizes the results from a number of investigations of this kind. Perhaps the most intriguing finding is that, while cis isomers of platinum complexes in both the 2+ and 4+ platinum oxidation states show high levels of activity, the corresponding trans isomers are ineffective as chemotherapeutic agents. This stereochemical specificity certainly stirs the curiosity of the coordination chemist. What structural details of the site of interaction require this selectivity? Does a biological ligand chelate the platinum reagent, a process requiring a cis orientation? Or is this specificity the result of subtle hydrogen bonding interactions such as found in the binding of *cis*- and *trans*-DDP to tRNA? Alternatively, does the cis configuration of the complex determine a sequence of kinetic reactions that are unfavorable in the case of the trans reagent? The relative reactivities of these square planar complexes have been long known to depend on the ligand orientation (170, 171). Much research effort has been directed toward resolving these chemical differences.

Figure 28 displays the general structural features that distinguish active from inactive platinum reagents in both the 2+ and 4+ platinum oxidation states. Since the bulk of clinical and chemical investigations have centered on *cis*-DDP, we focus our attention on the platinum(II) complexes. Diammineplatinum(II) compounds having ligands with moderately high leaving ability, such as chloride or bromide, or weakly bound bidentate ligands, such as malonate or oxalate, show antitumor activity (5, 167). Diammineplatinum(II) complexes possessing two loosely bound ligands such as nitrate anions have however proved to be too toxic. The nature of the amine groups also influences the selectivity of the drug. Activity increases along the series methylamine to *n*-butylamine. Secondary amines are less effective. Chelating ligands such as ethylenediamine lead to active drugs in some tumor systems, although of lower activity than the parent diammine complex. Compound solubility is also a factor in determining the efficacy of the platinum drug. Moreover, the relative activity of the drug is always dependent on the tumor system tested. The two complexes showing the highest and broadest spectrum of activity, and which presumably will be examined in second-generation clinical trials,

Table 8 Structure–Activity Studies of Platinum Antitumor Drugs

Compound	LD_{50}^a (mg/kg)	ID_{90}^b (mg/kg)	Thera-peu-tic[c,d] Index	Ref-erence
cis-(NH$_3$)PtCl$_2$	13.0	1.6	8.1	168
trans-(NH$_3$)$_2$PtCl$_2$	27.0	> 27.0	< 1.0	168
(en)PtCl$_2$	22.5	10	2.25	168
cis-(methylamine)$_2$PtCl$_2$	18.5	12	1.5	169
cis-(ethylamine)$_2$PtCl$_2$	26.5	12	2.2	169
cis-(*n*-propylamine)$_2$PtCl$_2$	26.5	12	2.2	169
cis-(*n*-pentylamine)$_2$PtCl$_2$	110	10	11	169
cis-(isopropylamine)$_2$PtCl$_2$	33.5	0.9	37.1	169
cis-(isobutylamine)$_2$PtCl$_2$	83	6.2	13.4	169
cis-(cyclopropylamine)$_2$PtCl$_2$	56.5	2.3	24.6	168
trans-(cyclopropylamine)$_2$PtCl$_2$	27	> 27	—	168
cis-(cyclobutylamine)$_2$PtCl$_2$	67	< 6	> 11.1	168
cis-(cyclopentylamine)$_2$PtCl$_2$	480	2.4	200	168
cis-(cyclohexylamine)$_2$PtCl$_2$	> 3200	12	> 267	168
Platinum(IV) compounds				
cis(Cl)/*trans*(OH)-(NH$_3$)$_2$Pt(OH)$_2$Cl$_2$	135	4.8	28.1	169
cis-(isopropylamine)$_2$PtCl$_4$	111	< 10	> 11.1	169
cis-(cyclopentylamine)$_2$PtCl$_4$	141	3.0	48	169

Variations in Labile Ligand	Toxic Level[e]	$T/C^{f,g}$	Reference
cis-(NH$_3$)$_2$PtCl$_2$	9	1	167
cis-(NH$_3$)$_2$Pt(NO$_3$)$_2$	7	54	167
cis-(NH$_3$)$_2$PtBr$_2$	5–6	13	167
cis-(NH$_3$)$_2$PtI$_2$	> 25	110	167
cis-(NH$_3$)$_2$Pt(SCN)$_2$	~ 50	70	167
cis-(NH$_3$)$_2$Pt(NO$_2$)$_3$	> 400	99	167
cis-(NH$_3$)$_2$Pt(oxalate)	35	73	167
cis-(NH$_3$)$_2$Pt(malonate)	45–60	18	167

[a]LD_{50} is a measure of toxicity, the lethal dose for 50% of the animal population.
[b]ID_{90} is a measure of potency, the inhibiting dose at which 90% of the tumor cells are killed.
[c]TI is the therapeutic index, which equals LD_{50}/ID_{90}.
[d]Assays were conducted using the ADJ/PC6A tumor system in mice.
[e]The toxic level is defined as that dosage which is lethal to 83% of the animal population.
[f]For solid tumors, inhibition is measured in terms of the ratio of weights of treated and untreated (control) tumors, T/C. Values less than 50 are generally considered significant.
[g]Assays were conducted on solid Sarcoma 180 tumors transplanted in mice.

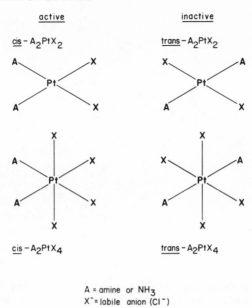

Figure 28 Active and inactive antitumor platinum complexes.

are dichlorodicyclopropylamineplatinum(II) (168) and 1,2-diaminocyclo-hexaneplatinum(II) sulfate (172).

Activity appears to require that the platinum coordination sphere contain two cis-oriented labile ligands, such as chloride, and two substitutionally inert amine ligands. In aqueous solution these labile ligands may be easily displaced, leaving two highly reactive cis-oriented sites for ligand coordination. With some certainty, it appears that the cis-$(NH_3)_2Pt^{2+}$ moiety binds the biological target molecule.

The hydrolytic equilibria (173) of platinum(II) complexes suggest the general scheme depicted in Figure 29. In the plasma, the high concentration of chloride ion (0.103 M) suppresses ligand dissociation. The neutral platinum drug therefore remains in an unreactive, nontoxic form outside the cell. After passive diffusion of the neutral complex across the cell membrane, the drug becomes activated by the loss of chloride ion. Within the cytoplasm, the chloride ion concentration is approximately 4 mM. The resultant cationic diammineplatinum(II) hydrolysis products, among them $[(NH_3)_2Pt(OH_2)(OH)]^+$, should bind readily to cellular targets such as the polyanionic DNA. Since water is an excellent leaving group, the platinum drug is converted to a highly reactive species. It is generally argued that cis-DDP subtly manifests its activity through this solvent-assisted pathway.

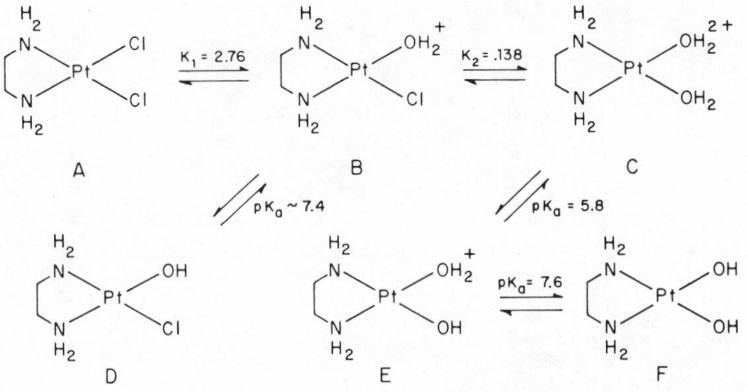

Estimate of Species A–F in a Biological System, pH = 7.4

	[Cl⁻]	A	B	C	D	E	F
Plasma	103 mM	37.3	1	.00134	1	.053	.033
Cytoplasm	4 mM	1.45	1	.0345	1	1.38	0.86

Figure 29 Hydrolysis reactions of antitumor platinum complexes and an estimate of the species present in the plasma and cytoplasm. [Reproduced with permission from ref. 215.]

4.1.2 DNA as the Cellular Target.

While it remains to be established definitively that *cis*-DDP exerts its antitumor effects through interaction with DNA, the overwhelming evidence thus far points to DNA as the cellular target. The first observation of biological activity of this drug, upon which the possible anticancer activity was suggested, was that the presence of *cis*-DDP leads to the filamentous growth of bacteria; *cis*-DDP effectively inhibits the division of bacterial cells without the concomitant inhibition of cellular growth. It appeared, then, that the agent could selectively inhibit DNA synthesis with no accompanying effect on RNA or protein synthesis. In the presence of other DNA-specific antitumor agents or as a result of X-ray irradiation, this same cellular phenomenon is observed.

Selective inhibition of DNA synthesis has been demonstrated in both in vivo (174) and in vitro (175) systems. By measuring the incorporation of [³H]-thymidine, [³H]-uridine, and [¹⁴C]-leucine into tumor cells treated with *cis*-DDP, the rates of DNA, RNA, and protein synthesis, respectively, were determined. At low drug concentrations DNA synthesis in human amnion AV3 cells was selectively inhibited, while higher drug concentrations resulted in the inhibition of all three processes. Similarly, after intraperitoneal injection of *cis*-DDP into mice bearing developed tumors, in vitro assessment of the precursor incorporations into the tumor cells revealed initial suppression of the rates of synthesis of DNA, RNA, and protein but the persistent suppression only of DNA synthesis. These data point to DNA as the primary cellular target of the drug.

Simplistically, then, the anticancer activity of the drug could be viewed as the direct result of the interaction of the platinum complex with the DNA. In binding to the DNA, *cis*-DDP could block replication. This inhibition of DNA synthesis would suppress the rapid growth of cancerous tissue. But why does the drug selectively affect tumorigenic cells at low doses? *Cis*-DDP appears to kill specifically cancerous cells with comparatively low damage to normal tissue. It is difficult to reconcile the selectivity of this agent based solely on the premise that cancerous cells, dividing more readily, will have an increased uptake of the drug and therefore more efficient inhibition of DNA synthesis. Tissue distribution studies (176, 177) in fact show higher concentrations of platinum in normal tissues such as the kidney, liver, and spleen than in the tumor. An alternative mechanism for specificity must therefore be invoked. Moreover, is the effect the direct consequence of platinum binding to the DNA? The drug could block DNA synthesis through binding to an enzyme involved in replication and inhibiting the action of this enzyme.

As is also found with chemotherapeutic DNA-alkylating agents, *cis*-DDP induces the lysogenic activity of bacteria. The following experiment (178) strongly indicates the primary site of interaction to be the DNA. *Escherichia coli* cells were infected with a lysogenic phage, with the incorporation of the viral genome into the host DNA. This infected strain, denoted F+, remained in a latent state, growing and dividing but without the expression of the viral DNA. *Cis*-DDP was then added to an F− strain which had not been infected with virus. Following the conjugation of F+ and F− strains, lysis of the F+ cells was observed. This experiment suggests two important conclusions. First, plasmid DNA from the F− cell appeared to act as a carrier for the platinum complex to the F+ cell. The platinum reagent had bound to cellular DNA. Secondly, the action of the platinum complex had derepressed the viral gene in the F+ host; expression of the viral DNA resulted in the subsequent lysis of the cell. Binding of the platinum complex to the recombined host DNA could effect such a change.

The strong correlation between lysogenic induction, anticancer activity, and, paradoxically, carcinogenicity for a number of chemotherapeutic agents, among them *cis*-DDP, has led to speculations that *cis*-DDP may exert its effects through an immunosuppressive mechanism (179). By derepressing the transformed genome in a manner similar to lysogenic induction, the antigenicity of the cancerous cell could be expressed and therefore made susceptible to the immune system of the body. The details of this biological mechanism are considered in Chapter 1. This interesting hypothesis could explain the preferentially lethal effects of the antitumor drug on cancerous rather than normal tissue.

The interaction of *cis*-DDP with DNA leads, as one might expect, to the induction of mutations. As a result of numerous genetic assays in bacterial systems, *cis*-DDP has been shown (180, 181) to be a potent base substitution mutagen. The analogous trans isomer is a substantially less effective mutagen and does not alter the DNA through a "base substitution" mechanism. Moreover, *cis*-DDP is exceptionally lethal to DNA repair-deficient mutants (182, 183). An alternate proposal for the selective action of *cis*-DDP in tumorigenic cells is based on the relative efficiencies with which cancerous versus normal cells could repair these platinum-induced lesions in the DNA (184–186).

4.2   The Differential Biochemical Effects of *cis*-and *trans*-DDP

Assuming the primary cellular target of *cis*-DDP to be the DNA, a number of questions arise. What are the biochemical effects of the lesion introduced in the DNA as a result of platinum binding, and can a difference in effects between the active cis and inactive trans isomers be detected? Moreover, does the requirement for the cis isomer reflect a specific lesion in the DNA induced by this platinum complex that can be correlated with cytotoxic activity?

4.2.1   Cell Survival. An important correlation that was determined (187) in the cases of both *cis*- and *trans*-DDP was the relationship between cell survival and the amount of drug bound to the DNA. The effects of *cis*- and *trans*-DDP on the colony-forming abilities of HeLa cells were measured and compared with the level of platinum binding to the DNA extracted from cells that had been treated with platinum at similar doses; in these experiments platinum binding was assayed by atomic absorption spectroscopy. As indicated in Figure 30, higher levels of the trans isomer were required to bind to the DNA in order to effect the same reduction in cell survival as the cis isomer. From the data it is evident that the differences in activities of *cis*- and *trans*-DDP cannot result solely from differences in membrane permeabilities. There must instead be a real difference in the reaction of these complexes with DNA upon which the differential cytotoxic effects are based. Approximate calculations using these data show that 45 *cis*-DDP molecules or 170 *trans*-DDP molecules are bound to each DNA molecule (assuming MW $\sim 10^9$) for a measurable reduction in cell survival. In addition, comparison of these results with levels of binding to other cellular components again supports the contention that DNA is the primary target of the platinum(II) compounds.

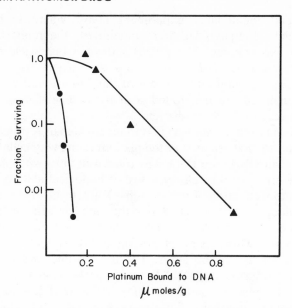

Figure 30 Relationship between the survival of HeLa cells treated with *cis*-DDP(●) and *trans*-DDP(▲) and the extent of platination of the DNA isolated from parallel cultures (187).

4.2.2 Replication and Transformation. In a variety of in vitro assays, both *cis*- and *trans*-DDP have been shown to inhibit replication. Experiments (188) monitoring the incorporation of nucleotide precursors into acid-precipitable counts, reflecting polymer formation, after pretreatment of the nucleic acid template with platinum reagent indicated that *cis*-DDP inactivates templates for human DNA polymerase and Rausher murine leukemia virus reverse transcriptase. In an extension (189) of this work, both a single-stranded DNA having a heterogeneous sequence and the synthetic copolymers poly(dA)·poly(dT) and poly(dAT) were examined as template primers. Levels of platinum binding were assayed by atomic absorption spectroscopy. To achieve 50% inhibition, a lower level of platinum binding was required using poly (dAT) than the single-stranded DNA as template primer, suggesting some sequence specificity to the inhibition. Regardless of the primer employed, the trans isomer was less effective in inhibiting the synthesis of DNA. To assure that platinum released from the template and then bound to the polymerase did not cause the inhibitory effect, platinated templates were chromatographed before enzyme addition. Moreover, enzyme preincubated with *cis*-DDP showed inhibition only at high doses of platinum, again indicating the primary target site to be the DNA.

In a recent double-label experiment (190), where both tritiated thymidine and [^{195m}P]cis-DDP were monitored, the replication of T7 phage DNA was assayed. The results indicated that while *trans*-DDP inactivates replication, *cis*-DDP does so at an order of magnitude lower level of binding. In addition, the rate of inhibition found with *cis*-DDP paralleled quantitatively that found with pyrimidine dimers which had formed as a result of UV irradiation.

Cis-DDP has also been shown (191) to inactivate DNA transformation. Again in order to limit reaction of the platinum drug with cellular components other than DNA, the DNA was treated in vitro with *cis*-DDP and thereafter returned to the cell for assay of biological activity. The data indicated that inactivation of transforming DNA results from an interference in the integration of the DNA into the recipient genome; integration of platinum-bound transforming DNA decreases with an increase in platinum binding. This loss of integrating ability parallels the decreased transforming activity of the DNA. The trans isomer induces a similar inactivation of transforming DNA but with a 50% decreased efficiency.

In summary, then, *cis*- and *trans*-DDP inhibit nucleic acid processes; but in all systems studied thus far, effects of the active antitumor drug are more pronounced. It is unclear whether this differential inactivation is the result of distinct differences in the lesion in the DNA induced by the cis and trans isomers or, alternatively, the result of different degrees to which these complexes induce the same lesion.

4.2.3. Crosslinking. The primary lesion in the DNA induced by the platinum reagent was initially thought to be the result of the formation of interstrand crosslinks. The requirement for a cis orientation of labile ligands suggests that the drug binds the DNA in a bifunctional manner, with ligand coordination through two sites in the platinum coordination plane. Moreover, the spectrum of activity of *cis*-DDP and the effects of *cis*-DDP on lysogenic bacterial growth and cellular DNA synthesis resemble those of classical bifunctional alkylating agents which are believed to crosslink DNA strands. This proposal would explain the low levels of platinum binding required for activity; crosslinking of the DNA strands would prevent the strand separation necessary for subsequent replication.

Crosslinking of complimentary strands of DNA by *cis*-DDP has been demonstrated (192). In a typical experiment to detect crosslinking, HeLa cells were grown in a medium containing a density label, 5-bromo-2'-deoxyuridine (BrdU). After incorporation of this heavy label into one strand of the cellular DNA, extraction and subsequent alkaline cesium chloride gradient centrifugation (denaturing conditions) resulted in two bands consisting of "heavy" and "light" DNA strands. Crosslinking of

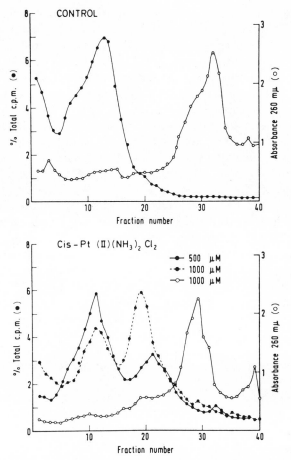

Figure 31 Formation of crosslinked DNA in HeLa cells treated with *cis*-(NH$_3$)$_2$PtCl$_2$. Depicted are the alkaline cesium chloride gradient profiles of DNA isolated from cells either not treated (control) or treated with *cis*-(NH$_3$)$_2$PtCl$_2$. Cells were grown in a medium containing both a radioactive (^{3}H-dT) and a density (Br-dU) label for the DNA. As the control profile indicates, tritiated thymidine is incorporated only into the heavy strand. The light DNA is assayed less sensitively by its ultraviolet absorption (A_{260}). After treatment with platinum at dosages of 500 μM (•———•) or 1000 μM (• - - -•), a labeled peak at intermediate density appears corresponding to the crosslinked strand. [Reproduced by permission from ref. 192].

the DNA would produce an intermediate "hybrid" species. As shown in Figure 31, after cell incubation with *cis*-DDP, crosslinking of the DNA strands could be detected.

The question remained whether the extent of crosslinking could be correlated with the relative activities of platinum(II) complexes and, further, whether the frequency of crosslinking was sufficient to account for the cytotoxic effects of *cis*-DDP. In an extension of the work cited above, crosslinking of DNA by both *cis*- and *trans*-DDP was measured (187) in an in vivo assay, where cells were incubated with platinum reagent, and an in vitro assay, where isolated DNA was incubated with platinum reagent. While the cis and trans isomers showed comparable abilities (1.5 to 2.0 greater frequency for *cis*-DDP) to crosslink the DNA in vitro, in vivo assessment revealed a 10-fold greater frequency of crosslinks in the case of the cis isomer for a given dosage of platinum. The frequency of crosslinking for *cis*-DDP calculated from these in vivo results, however, indicated that only 1 in 400 platination reactions resulted in a crosslink as compared with a frequency of 1 in 8 calculated for the alkylating agent, mustard gas.

A study (193) designed to correlate crosslinking with the inactivation of T7 bacteriophage further indicated that interstrand crosslinking reactions were unlikely to be the primary cytotoxic lesion. At a dosage of platinum necessary to inactivate all the phage particles, only a small percentage of the phage contained crosslinked DNA. Similarly, in the study (191) assaying the inactivation of transforming DNA by *cis*-DDP, it was found that while 1 platinum molecule per 1000 bases was the biologically inactivating dose, a crosslink is formed when there are as many as 12 platinum molecules bound per 1000 bases.

4.3   Chemical Features of the Interaction of DDP with DNA

Of prime importance in determining the mode of action of *cis*-DDP is the elucidation of the site of interaction between the antitumor drug and the DNA molecule. Based on studies of the effects of platinum binding on the structure of DNA, some proposals have been put forth. As we shall see, the techniques discussed previously (Section 3.2) can be utilized to gain some insights into the reactivity of DDP with the native DNA duplex. The examination of the interaction of *cis*- and *trans*-DDP with nucleic acid components and synthetic DNAs having specific sequences has furthermore led to a number of structural models for the lesion introduced in the DNA by *cis*-DDP. These models depend in part on the differential activities of the cis and trans isomers.

4.3.1   Binding Properties and Effects on DNA Structure.   Both the kinetic and thermodynamic parameters associated with the reaction of DDP with DNA have been studied. *Cis*- and *trans*-DDP bind to DNA with similar kinetics (194). Studies of the binding of *cis*- and *trans*-DDP to T7 phage DNA show the rates of association to be sensitive functions of the buffer system employed. In solution, buffering anions may displace the bound chloride ions and, depending on the leaving group ability of these anions, may increase or decrease the rate of binding to DNA. The comparatively slower binding rates of the *trans*-DDP complex may be a function of the lower solvolysis rate of this isomer. Recently, oligomeric hydrolysis products of *cis*-DDP have been isolated and characterized. The planar hydroxo-bridged dimer (195) of *cis*-diammineplatinum(II) and analogous trimer (196) may be important intermediates in the intracellular reactions of *cis*-DDP. Equilibrium dialysis studies (197) of the interaction of $[^{14}C][(en)PtCl_2]$ with *E. coli* DNA indicated that $(en)PtCl_2$ binds reversibly to the DNA with a saturation ratio of 0.57 Pt : DNA-phosphate. This result is consistent with the antitumor analog being bound to the DNA in both monofunctional and bifunctional modes. These studies carried out at high levels of binding however, while indicative of various nonspecific modes of association, do not necessarily reflect the specific binding of the platinum reagent to the DNA at the low ratios (r) of platinum : DNA-phosphate present within the cell.

Cis-DDP has been shown (159) to bind specifically to DNA rich in guanine–cytosine content. Binding of this heavy metal complex to DNA results in large increases in the polymer buoyant density. Following the incubation of *cis*-DDP with mixtures of radiolabeled DNAs of different guanine–cytosine contents, a separation of the platinated DNAs was conducted by cesium chloride gradient centrifugation. The buoyant densities, and therefore the levels of platinum binding, of these DNAs were found to increase linearly with the guanine–cytosine content of the DNA. At the very low r values within the cell, a tight association of the platinum drug with a specific guanine–cytosine-rich DNA sequence becomes quite plausible.

It has been established that the square planar platinum complex binds covalently to the DNA rather than in a noncovalent intercalative fashion. *Cis*-DDP has been shown (128, 198) to inhibit noncompetitively the fluorescence of ethidium bromide intercalated into duplex DNA (see Section 3.2). The bound platinum reagent must therefore alter the DNA structure so as to block the intercalation of the ethidium cation between the base pairs; local denaturation with the disruption of base-pairing through inter- or intrastrand crosslinking, for example, could cause this noncompetitive inhibition. The possibility that *cis*-DDP depurinates DNA, suggested by analogy to a suspected mode of action of alkylating reagents (199), has not yet been experimentally tested.

What is known about the effects of this covalent binding of *cis*- and *trans*-DDP on the structure of duplex DNA? Raman spectra (52) monitoring changes in the sugar–phosphate backbone vibrational modes reveal a disruption of the native B-form structure of the calf thymus DNA helix on binding *cis*- or *trans*-DDP. The effects of *cis*- and *trans*-DDP on the thermal melting of calf thymus DNA have also been studied (200). At low r values, *cis*-DDP appears to increase the T_m slightly, while higher levels of binding of the cis isomer cause a destabilization of the helical form. Only slight changes in T_m were observed in the case of the trans isomer, although it is unclear whether, under the conditions employed, significant concentrations of *trans*-DDP were bound to the DNA. All platinum complexes studied (cis and trans analogs of DDP in the 2+ and 4+ oxidation states) were found to renature the DNA at low r values, on which basis a stabilizing crosslink of the DNA was suggested.

The hydrodynamic properties of the DNA–platinum complex suggest a significant distortion of the regular helical duplex structure. Sedimentation studies (201) reveal a large increase in the sedimentation coefficient of T7 phage DNA on binding either *cis*-DDP, *trans*-DDP, or (en)PtCl$_2$. The increase in sedimentation rate of the native DNA as a function of binding *cis*-DDP and (en)PtCl$_2$ was found to be greater than would be expected based on estimated changes in mass and density. This result indicates a conformational alteration of the DNA structure induced by the platinum reagent. The authors have suggested that local regions of denaturation which would collapse the DNA into a compact structure on sedimenting would best account for the data. The trans isomer, while increasing the rate of sedimentation, does so to a relatively small extent for a given degree of platinum binding. Estimates of the extent of binding of the trans reagent were made only on the basis of bouyant density shifts, however. Viscosity measurements (202) also support a local denaturation of the duplex. The viscosity of the duplex DNA decreases sharply with low levels of platinum binding, $r \leqslant 0.05$. The fraction of unpaired bases induced by the binding of *cis*-DDP was assayed using a formaldehyde kinetic method and, based on this assay, appeared too few to account for the dramatic decrease in viscosity observed. A specific localized denaturation or "kink" has been proposed to explain these results.

Finally, the binding of *cis*- and *trans*-DDP to supercoiled DNA has been studied (203) and provides dramatic evidence that platinum binding induces an unusual local denaturation of the helix. Figure 32 shows the gel electrophoretic patterns of closed and nicked PSM1 DNAs that had been incubated with *cis*- and *trans*-DDP, as a function of time. An increase in the electrophoretic mobility of the nicked circular DNA is observed, indicating the formation of a compact structure upon binding. More striking however is the change in the mobility of the closed circular DNA. The supercoiled DNA appears to unwind as a function of platinum binding. The mobility decreases as the supercoiled structure unfolds until it reaches a minimum coincident with the mobility of the nicked form,

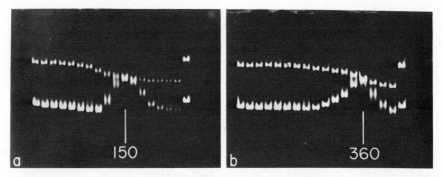

Figure 32 Electrophoresis (203) in 1% agarose gels of nicked and closed circular PSM1 DNA incubated with (a) *cis-* and (b) *trans*-$(NH_3)_2PtCl_2$ as function of time. After electrophoresis, gels were stained with ethidium bromide.

which has no superhelical turns. As the duplex continues to unwind into a negative supercoil, the mobility again increases. While characteristically similar to the effects of intercalative binding, these results must reflect instead a conformational change in the superhelix brought about by the covalent binding of the platinum complex. Local disruptions of the base pairs could result in this unwinding. As is evident, both the cis and trans isomers induce this conformational change. In fact, assays of bound platinum by atomic absorption spectroscopy reveal a similar dependence of electrophoretic mobilities of the DNA on bound platinum concentration. These results further suggest a shortening of the DNA with platinum binding, and electron micrographs show the platinated DNA to be shortened by as much as 50% of its original length. Binding of these platinum complexes causes a collapsing of the DNA structure.

The studies cited above resolve a number of questions but raise others. *Cis*-DDP and *trans*-DDP bind covalently to the DNA duplex, inducing a conformational distortion in the structure. This disordering could be the result of the opening of the bases, disrupting base-pair hydrogen bonds to accommodate platinum binding. These complexes appear to denature localized regions of the duplex that may have a high guanine–cytosine content. But what are the structural details of this site of interaction? Moreover, a distinction between the binding of cis and trans isomers remains to be established. In few studies thus far have significant differences been noted. Instead, only a difference in the extent of binding necessary to cause a particular effect has been observed, and this difference may be a function only of the experimental design employed to assay for platinum binding. What differences in the reactivities of these isomers determine their differential cytotoxic activities?

4.3.2   Models for the Site of Interaction.   A variety of structural models for the site of interaction of *cis*-DDP and DNA have been proposed. Both solution and crystallographic studies of the binding of platinum(II) complexes to nucleic acid constituents (as described in Section 2) establish the primary site of coordination to the platinum atom to be through the N-7 nitrogen atom of purines and the N-3 nitrogen atom of pyrimidines. This simple, monofunctional ligand coordination however is not sufficient to distinguish the antitumor activities of cis and trans isomers. In addition, potentiometric studies (204) indicate the release of both chloride anions on binding of *cis*-DDP to DNA; *trans*-DDP, in contrast, appears to release one chloride in binding to the DNA. The *cis*-diammineplatinum(II) moiety binds to the DNA through two sites in the platinum coordination plane.

Consider the possible bifunctional modes of coordination. The cis-oriented platinum coordination site could be occupied by two ligating atoms from one base residue. The "N-7–O-6" proposal, considered below, exemplifies this mode. The trans isomer could not bind in an equivalent fashion. Alternatively, the *cis*-diammineplatinum(II) complex may bind both the base and phosphate moieties of a particular nucleotide unit. Other possible bifunctional modes include the formation of either an inter- or intrastrand crosslink. While unlikely in view of the evidence cited previously, the crosslinking of base residues on opposite strands of the duplex remains as a viable proposal; this interstrand crosslink would severely hinder replication. At present, it appears more probable that the site of interaction involves an intrastrand crosslink. The observation that the nonbonded distance separating chlorine atoms in *cis*-DDP is 3.4 Å, the base-pair stacking distance, prompted numerous proposals for this intrastrand crosslink. Bridging of stacked base pairs would result in distances shorter than 3.4 Å between nitrogen and/or oxygen ligating atoms however and would necessitate the introduction of bent bonds. There are in fact many chemically feasible and interesting proposals for, and examples in characterized mononucleotide–platinum complexes consistent with, intrastrand crosslinking. Various of the above structural models are depicted in Figure 33.

One proposal for the mode of action of *cis*-DDP that has provoked heated debate is that a chelate complex of *cis*-diammineplatinum(II) forms with the guanine base through N-7 and O-6 linkages (205–207) in a fashion analogous to the 6-mercaptopurine–palladium complex cited previously (Fig. 7). This model (Fig. 33*a*) is consistent with the observation of G–C specificity, and certainly such a complex would result in a base-substitution mutation. Perturbations of the C$=$O stretching frequencies in the infrared spectra of complexes isolated from solutions containing *cis*-DDP and guanosine have been noted, although these complexes have not

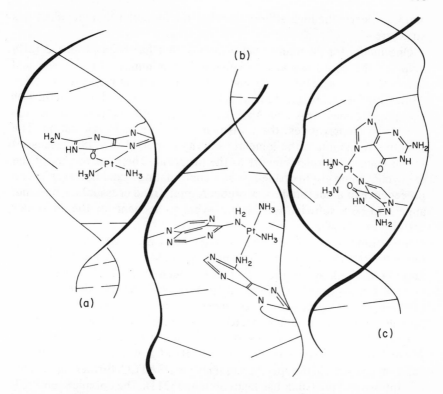

Figure 33 Proposals for the site of interaction of *cis*-(NH₃)₂PtCl₂ with the DNA duplex: (a) chelation of guanine bases; (b) an interstrand crosslink between adenine units; (c) an intrastrand crosslink between adjacent guanine bases. [Reproduced with permission from ref. 215.]

been structurally characterized. Raman spectra (52) of solutions containing *cis*-DDP and guanosine do not reveal comparable perturbations of the ketone stretching modes. X-Ray photoelectron spectra of samples containing calf thymus DNA incubated with *cis*-DDP show shifts in the oxygen 1*s* binding energies; these shifts may reflect metal coordination to atoms other than O-6 of the guanine base, however. Spectra of *cis*-DDP complexes with guanine were not reported. No direct evidence has substantiated the "N-7–O-6" proposal. To the contrary, many authors (52, 64, 208, 209) have argued against this mode of chelation on stereochemical grounds. Projections from the lone pairs of electrons associated with the carbonyl oxygen O-6 and ring nitrogen N-7 intersect at an angle of approximately 37°. Simultaneous binding of the metal to both centers would produce considerable strain energy. The strain in the 6-mercaptopurine–palladium complex is manifest by angular distortions of the purine ring, and it is clearly the high affinity of palladium for sulfur that stabilizes this structure.

One model for an interstrand crosslink that has been proposed (210) involves the coordination of *cis*-diammineplatinum(II) to opposite and adjacent adenine residues through the N-6 amino nitrogen atoms (Fig. 33*b*). Circular dichroism studies (211) of *cis*- and *trans*-DDP reaction with the dinucleotides ApA and ApC reveal that, while the trans isomer unstacks the dinucleotides, the cis isomer stabilizes their association. The selective reduction in the template activity of poly(dAT) (189) discussed above lends additional support to the proposal. The involvement of the exocyclic amino group of adenine as a coordinating ligand however has no precedence in model studies. Moreover, interstrand crosslinking does not appear to be a sufficiently frequent event to account for the antitumor activity of *cis*-DDP.

Intrastrand crosslinking is the most appealing proposal. The base tilting required to accomplish this crosslinking would lead to the local denaturation or "kinking" of the helix that has been observed. Given the preferential binding of *cis*-DDP to GC-rich DNAs, the added observation (212) that DDP produces large increases in the buoyant density of poly(dG)·poly(dC) and unexpectedly small shifts in the alternating copolymer, poly(dGC), provides strong evidence in support of this proposal. Using the restricted endonuclease BAM-1, which cleaves between adjacent guanines in the specific sequence G ↓ GATCC, further support for this intrastrand crosslink has been obtained (213). The complex (en)PtCl$_2$ alters the cleavage pattern of this endonuclease. If cyanide, a ligand having a high affinity for platinum, is subsequently added to the solution, the strands may be separated. The *cis*-diammineplatinum(II) moiety appears to hold the cleaved strands together until the cyanide ion is introduced.

This intrastrand crosslink could be accomplished in a number of ways. One possibility is the coordination of the platinum through the N-7 nitrogen atoms of neighboring guanine bases (Fig. 33*c*). Alternatively, the bridging of adjacent residues could be achieved in a fashion similar to that indicated by the structures of the oligomeric complexes of *cis*-diammineplatinum α-pyridone blue (68), *cis*-diammineplatinum-l-methyl thymine dimer (72), or the silver cytosine dimer (73) discussed previously. Two platinum atoms could serve to bridge neighboring guanine residues, as depicted in Figure 34, with coordination of one platinum to the deprotonated nitrogen atom, N-1, and coordination by the other platinum atom to the exocyclic oxygen atoms, O-6. Platinum binding to these base-paired hydrogen bonding positions would surely distort the helix locally, skewing the polymer backbone in order to bridge neighboring bases tightly. An

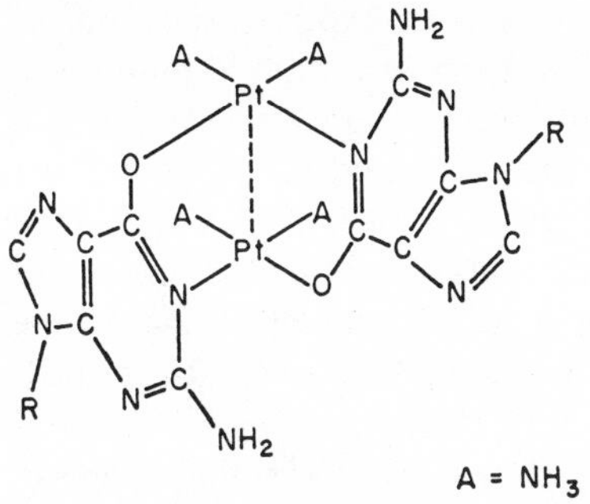

Figure 34 A possible complex between *cis*-diammineplatinum units and guanine bases. [Reproduced with permission from ref. 215.]

A = NH$_3$

EXAFS (Extended X-Ray Absorption Fine Structure) study (214), useful in determining the radial distance between heavy metals in amorphous solids or in solution, was aimed at detecting the short platinum–platinum distances this proposal suggests. No metal-metal distance shorter than 3 Å was observed in spectra of cis and trans complexes with calf thymus DNA. The spectra were not sufficiently sensitive however to detect slightly longer nonbonded distances between platinum(II) centers, which are possible in this structural arrangement. The spectra were consistent with the loss of chloride ions and the formation of four Pt–N or Pt–O linkages upon platinum binding to the DNA.

The reactivities of these oligomeric platinum-base complexes suggest one further structural possibility. *Cis*-diammineplatinum α-pyridone blue has been found to release ammonia with decomposition. In binding to the bases the *cis*-diammineplatinum(II) moiety could similarly release ammonia, leaving two additional reactive sites for ligand coordination (215, 216). This reaction would be promoted by the trans-labilizing effect of the coordinated bases. A highly stable, tightly associated complex with inter- and intrastrand crosslinks could result. Binding of *trans*-diammineplatinum(II) to a base position would labilize only the ligand, possibly still a chloride ion, coordinated to the site trans to it; no additional sites in the platinum coordination plane would become exposed. Mass-spectrometric studies (217) of the interaction of *cis*- and *trans*-DDP with nucleosides reveal the labilization of ammonia, although the authors have quite reasonably suggested the release to be a consequence of the high temperature conditions in the mass spectrometer. Efforts to detect ammonia release on platinum binding to DNA are currently under way.

A variety of additional structural proposals for the site of interaction have been put forth. One possibility is a DNA-protein crosslink (218, 219), which would block further strand replication. The high affinity of platinum for sulfur suggests the possibility of metalation at exposed cysteine or methionine residues of a polymerase. While there is a high likelihood of some protein binding within the cell, the in vivo and in vitro results point to DNA as the primary site of cytotoxicity. In binding to the DNA, platinum coordination however is not restricted to the base residues. The structure (83) of [Pt(en)(5'CMP)]$_2$ exemplifies covalent binding of platinum to the phosphate moiety. Binding to the phosphate backbone would result in severe alterations in the DNA conformation. This possible binding mode should not be overlooked.

The site of interaction of *cis*-DDP with DNA must certainly contain some of the structural features described above. A definitive elucidation of this site remains an exciting challenge for the future.

REFERENCES

1. (a) B. Rosenberg, L. Van Camp, J. E. Trosko, and V. H. Mansour, *Nature,* **222,** 385 (1969); (b) B. Rosenberg and L. Van Camp, *Cancer Res.,* **30,** 1799 (1970); (c) for a review, see J. M. Hill, E. Loeb, A. MacLellan, N. O. Hill, A. Khan, and J. J. King, *Cancer Chemother. Rep.,* **59,** 647 (1975).

2. D. J. Hodgson, *Prog. Inorg. Chem.,* **23,** 211 (1977).

3. L. G. Marzilli, *Prog. Inorg. Chem.,* **23,** 255 (1977).

4. A. J. Thomson, *Platinum Met. Rev.,* **21,** 2 (1977).

5. *J. Clin. Hematol. Oncol.,* **7** (1977).

6. R. W. Gellert and R. Bau, *Met. Ions Biol. Syst.,* **8,** 1 (1979).

7. V. Swaninathan and M. Sundaralingam, *CRC Crit. Rev. Biochem.,* 245 (1979).

8. T. R. Jack, *Met. Ions Biol. Syst.,* **8,** in press.

9. M. Sundaralingam, *Biopolymers,* **7,** 821 (1969).

10. S. Arnott, *Prog. Biophys. Mol. Biol.,* **21,** 265 (1970).

11. D. Voet and A. Rich, *Prog. Nucl. Acid Res. Mol. Biol.,* **10,** 183 (1970) and references cited therein.

12. S. Arnott, S. D. Dover, and A. J. Wonacott, *Acta Cryst.,* **B25,** 2192 (1969).

13. C. D. Jardetsky, *J. Am. Chem. Soc.,* **83,** 2919 (1961).

14. A. E. V. Haschemeyer and A. Rich, *J. Mol. Biol.,* **27,** 369 (1967).

15. G. T. Rogers and T. L. V. Ulbricht, *Biochem. Biophys. Res. Commun.,* **39,** 419 (1970).

16. W. Saenger, *J. Am. Chem. Soc.,* **93,** 3035 (1971).

17. D. Suck and W. Saenger, *Nature,* **235,** 333 (1972).

18. R. M. Izatt, J. J. Christensen, and J. H. Rytting, *Chem. Rev.,* **71,** 439 (1971).

19. T. O'Connor, C. Johnson, and W. M. Scovell, *Biochim. Biophys. Acta,* **447,** 484 (1976).

20. W. Cochran, *Acta Cryst.,* **4,** 81 (1951).

21. J. Kraut and L. H. Jensen, *Acta Cryst.,* **16,** 79 (1963).

22. M. Sundaralingam, *Acta Cryst.,* **21,** 495 (1966).

23. V. Markowski, G. R. Sullivan, and J. D. Roberts, *J. Am. Chem. Soc.,* **99,** 714 (1977).

24. S. Lewin, *J. Chem. Soc.*, 792 (1964).

25. K. Nakanishi, N. Suzuki, and F. Yamazaki, *Bull. Chem. Soc. Japan*, **34**, 53 (1961).

26. B. W. Roberts, J. B. Lambert, and J. D. Roberts, *J. Am. Chem. Soc.*, **87**, 5439 (1965).

27. D. D. Jordan, *The Chemistry of Nucleic Acids*, Butterworth, Washington, D. C., 1960.

28. L. Katz and S. Penman, *J. Mol. Biol.*, **15**, 220 (1966).

29. C. E. Bugg, J. M. Thomas, M. Sundaralingam, and S. T. Rao, *Biopolymers*, **10**, 175 (1971).

30. M. Cohn, *Q. Rev. Biophys.*, **3**, 61 (1970).

31. R. J. P. Williams, *Q. Rev. Chem. Soc.*, **24**, 331 (1970).

32. B. S. Cooperman, *Met. Ions Biol. Syst.*, **5**, 79 (1976).

33. M. S. Chen, *Inorg. Perspect. Biol. Med.*, **1**, 217 (1978).

34. P. Tanswell, J. M. Thornton, A. V. Korda, and R. J. P. Williams, *Eur. J. Biochem.*, **57**, 135 (1975).

35. H. Sigel, *J. Am. Chem. Soc.*, **97**, 3209 (1975).

36. A. T. Tu and M. J. Heller, *Met. Ions Biol. Syst.*, **1**, 1 (1974).

37. S. K. Podder, *J. Mag. Res.*, **15**, 254 (1974).

38. S. Mansy and R. S. Tobias, *J. Am. Chem. Soc.*, **96**, 6874 (1974).

39. C. Y. Wei, B. E. Fischer, and R. Bau, *J.C.S. Chem. Commun.*, 1053 (1978).

40. B. A. Cartwright, D. M. L. Goodgame, I. Jeeves, and A. C. Skapski, *Biochim. Biophys. Acta*, **477**, 195 (1977).

41. J. A. Stanko and S. Hollis, results quoted by M. J. Cleare in *Platinum Coordination Complexes in Chemotherapy*, T. A. Connors and J. J. Roberts, Eds., Springer-Verlag, New York, 1974, pp. 24–26, and private communication.

42. Y. Y. H. Chao and D. R. Kearns, *J. Am. Chem. Soc.*, **99**, 6425 (1977).

43. J. F. Conn, J. J. Kim, F. L. Suddath, P. Blattman, and A. Rich, *J. Am. Chem. Soc.*, **96**, 7152 (1974).

44. R. B. Simpson, *J. Am. Chem. Soc.*, **86**, 2060 (1964).

45. M. J. Clarke and H. Taube, *J. Am. Chem. Soc.*, **96**, 5413 (1974).

46. S. Mansy, B. Rosenberg, and A. J. Thomson, *J. Am. Chem. Soc.*, **95**, 1633 (1973).

47. W. M. Scovell and T. O'Connor, *J. Am. Chem. Soc.*, **99**, 120 (1977).

48. D. Nelson, P. Yeagle, T. Miller, and R. B. Martin, *Bioinorg. Chem.*, **5**, 353 (1976).

49. P. C. Kong and T. Theophanides, *Inorg. Chem.*, **13**, 1981 (1974).

50. G. Kotowycz and O. Suzuki, *Biochemistry*, **12**, 5325 (1973).

51. S. Mansy, T. E. Wood, J. C. Sprowles, and R. A. Tobias, *J. Am. Chem. Soc.*, **96**, 1762 (1974).

52. (a) G. Y. H. Chu, S. Mansy, R. E. Duncan, and R. S. Tobias, *J. Am. Chem. Soc.*, **100**, 593 (1978); (b) S. Mansy, G. Y. H. Chu, R. E. Duncan, and R. S. Tobias, *J. Am. Chem. Soc.*, **100**, 607 (1978).

53. M. C. Lim and R. B. Martin, *J. Inorg. Nucl. Chem.*, **38**, 1915 (1976).

54. S. Mansy and R. S. Tobias, *J.C.S. Chem. Commun.*, 957 (1974).

55. C. J. L. Lock, R. A. Speranzini, G. Turner, and J. Powell, *J. Am. Chem. Soc.*, **98**, 7865 (1976).

56. R. M. K. Dale, D. C. Livingston, and D. C. Ward, *Proc. Natl. Acad. Sci. USA*, **70**, 2238 (1973).

57. H. J. Krentzien, M. J. Clarke, and H. Taube, *Bioinorg. Chem.*, **4**, 143 (1975).

58. F. B. Daniel and E. J. Behrman, *J. Am. Chem. Soc.*, **97**, 7352 (1975).

59. (a) S. Neidle and D. I. Stuart, *Biochim. Biophys. Acta*, **418**, 226 (1976); (b) T. J. Kistenmacher, L. G. Marzilli, and M. Rossi, *Bioinorg. Chem.*, **6**, 347 (1976).

60. A. B. Robins, *Chem. Biol. Int.*, **6**, 35 (1973).

61. L. G. Marzilli and T. J. Kistenmacher, *Acc. Chem. Res.*, **10**, 146 (1977).

62. S. J. Lippard, *Acc. Chem. Res.*, **11**, 211 (1978).

63. K. W. Jennette, S. J. Lippard, and D. A. Ucko, *Biochim. Biophys. Acta*, **402**, 403 (1975).

64. H. I. Heitner and S. J. Lippard, *Inorg. Chem.*, **13**, 815 (1974).

65. (a) J. P. Davidson, P. J. Faber, R. G. Fischer, Jr., S. Mansy, H. J. Peresie, B. Rosenberg, and L. Van Camp. *Cancer Chemother. Rep.*, **59**, 287 (1975); (b) B. Rosenberg, *ibid.*, **59**, 589 (1975).

66. B. Lippert, *J. Clin. Hematol. Oncol.*, **7**, 26 (1977).

67. R. J. Speer, H. Ridgway, L. M. Hall, D. P. Stewart, K. E. Howe, D. Z. Lieberman, A. D. Newman, and J. M. Hill, *Cancer Chemother. Rep.*, **59**, 629 (1975).

68. (a) J. K. Barton, H. N. Rabinowitz, D. J. Szalda, and S. J. Lippard, *J. Am. Chem. Soc.*, **99**, 2827 (1977); (b) J. K. Barton, D. J. Szalda, H. N. Rabinowitz, J. V. Waszczak, and S. J. Lippard, *J. Am. Chem. Soc.*, **101**, 1434 (1979).

69. J. K. Barton, S. A. Best, S. J. Lippard, and R. A. Walton, *J. Am. Chem. Soc.*, **100**, 3785 (1978).

70. J. K. Barton, C. Caravana, and S. J. Lippard, *J. Am. Chem. Soc.*, in press.

71. B. K. Teo, K. Kijima, and R. Bau, *J. Am. Chem. Soc.*, **100**, 621 (1978).

72. C. J. L. Lock, H. J. Peresie, B. Rosenberg, and G. Turner, *J. Am. Chem. Soc.*, **100**, 3371 (1978).

73. (a) L. G. Marzilli, T. J. Kistenmacher, and M. Rossi, *J. Am. Chem. Soc.*, **99**, 2797 (1977).

74. R. Faggiani, B. Lippert, C. J. L. Lock, and R. A. Speranzini, submitted for publication, October, 1979.

75. D. J. Szalda, L. G. Marzilli, and T. J. Kistenmacher, *Biochem. Biophys. Res. Commun.*, **63**, 601 (1975).

76. D. J. Szalda, T. J. Kistenmacher, and L. G. Marzilli, *J. Am. Chem. Soc.*, **98**, 8371 (1976).

77. J. A. Carrabine and M. Sundaralingam, *Biochemistry*, **10**, 292 (1971).

78. T. J. Kistenmacher, L. G. Marzilli, and D. J. Szalda, *Acta Crystalloys.*, **B32**, 186 (1976).

79. R. W. Gellert and R. Bau, *J. Am. Chem. Soc.*, **97**, 7379 (1975).

80. R. E. Cramer and P. L. Dahlstrom, *J. Clin. Hematol. Oncol.*, **7**, 330 (1977).

81. Y.-S. Wong and S. J. Lippard, *J.C.S. Chem. Commun.*, **22**, 824 (1977).

82. A. H. J. Wang, J. Nathans, G. van der Marel, J. H. Van Boom, and A. Rich, *Nature*, **276**, 471 (1978).

83. S. Louie and R. Bau, *J. Am. Chem. Soc.*, **99**, 3874 (1977).

84. K. Aoki, *Biophys. Acta*, **447**, 379 (1976).

85. D. M. L. Goodgame, I. Jeeves, C. D. Reynolds, and A. C. Skapski, *Nucleic Acids Res.*, **2**, 1375 (1975).

86. (a) K. Aoki, G. R. Clark, and J. D. Orbell, *Biochim. Biophys. Acta,* **425,** 369 (1976); (b) E. Sletten and B. Lie, *Acta Crystalloys.,* **B32,** 3301 (1976).

87. J. D. Watson and F. H. Crick, *Nature,* **171,** 964 (1953).

88. (a) V. Sasisekharan, N. Pattabiraman, and G. Gupta, *Proc. Natl. Acad. Sci. USA,* **75,** 4092 (1978); (b) V. Sasisekharan and N. Pattabiraman, *Curr. Sci.,* **45,** 779 (1976).

89. V. A. Bloomfield, D. M. Crothers, and I. Tinoco, Jr., *Physical Chemistry of Nucleic Acids,* Harper and Row, New York, 1974.

90. C. Tanford, *Physical Chemistry of Macromolecules,* Wiley, New York, 1961.

91. K. E. Van Holde, *Physical Biochemistry,* Prentice–Hall, Englewood Cliffs, New Jersey, 1971.

92. S. Arnott, "DNA Secondary Structures in Organization and Expression of Chromosomes," in Dahlem Konferenzen, Berlin, V. G. Allfrey, E. K. F. Bautz, B. J. McCarthy, R. T. Schimke, and A. Tissieres, Eds., 1976.

93. R. Langridge, H. R. Wilson, C. W. Hooper, M. H. F. Wilkins, and L. D. Hamilton, *J. Mol. Biol.,* **2,** 19 (1960).

94. R. Langridge, D. A. Marvin, W. E. Seeds, H. R. Wilson, C. W. Hooper, M. H. F. Wilkins, and L. D. Hamilton, *J. Mol. Biol.,* **2,** 38 (1960).

95. D. A. Marvin, M. Spencer, M. H. F. Wilkins, and L. D. Hamilton, *J. Mol. Biol.,* **12,** 60 (1965).

96. S. Arnott, D. W. L. Hukins, S. D. Dover, W. Fuller, and A. R. Hodgson, *J. Mol. Biol.,* **81,** 107 (1973).

97. S. Arnott, W. Fuller, A. Hodgson, and I. Prutton, *Nature,* **220,** 561 (1968).

98. S. Arnott, R. Chandrasekaran, D. W. L. Hukins, P. J. C. Smith, and L. Watts, *J. Mol. Biol.,* **88,** 523 (1974).

99. S. Arnott and P. J. Bond, *Nature, New Biol.,* **244,** 99 (1973).

100. J. L. Sussman and S. H. Kim, *Science,* **192,** 853 (1976).

101. A. Jack, J. Ladner, and A. Klug, *J. Mol. Biol.,* **108,** 619 (1976).

102. J. Vinograd, J. Lebowitz, R. Radleff, and P. Laipis, *Proc. Natl. Acad. Sci. USA,* **53,** 1104 (1965).

103. J. C. Wang, *J. Mol. Biol.,* **55,** 523 (1971).

104. J. Vinograd and J. Lebowitz, *J. Gen. Physiol.,* **49,** 103 (1966).

105. W. Keller, *Proc. Natl. Acad. Sci. USA,* **72,** 2550 (1975).

106. R. T. Espejo and J. Lebowitz, *Anal. Biochem.,* **72,** 95 (1976).

107. G. Felsenfeld, *Nature,* **271,** 115 (1978) and ref. cited therein.

108. J. T. Finch, L. C. Lutter, D. Rhodes, R. S. Brown, B. Rushton, M. Levitt, and A. Klug, *Nature,* **269,** 29 (1977).

109. H. Treibel and K. E. Reinhert, *Studia Biophys.,* **10,** 57 (1968).

110. P. D. Ross and R. L. Scruggs, *Biopolymers,* **6,** 1005 (1968).

111. J. Marmur and P. Doty, *J. Mol. Biol.,* **5,** 109 (1962).

112. T. Record, C. Woodbury, and T. Lohman, *Biopolymers,* **15,** 893 (1976).

113. G. L. Eichhorn and Y. A. Shin, *J. Am. Chem. Soc.,* **90,** 7323 (1968).

114. G. L. Eichhorn and P. Clark, *Proc. Natl. Acad. Sci. USA,* **53,** 586 (1965).

115. Y. A. Shin and G. L. Eichhorn, *Biochemistry,* **7,** 1026 (1968).

116. G. L. Eichhorn, *Nature,* **194,** 474 (1962).

117. C. K. S. Pillai and U. S. Nandi, *Biopolymers,* **12,** 1431 (1973).

118. T. Yamane and N. Davidson, *Biochim. Biophys. Acta,* **55,** 609 (1962).

119. R. Jensen and N. Davidson, *Biopolymers,* **4,** 17 (1966).

120. S. Katz, *J. Am. Chem. Soc.,* **74,** 2238 (1952).

121. T. Yamane and N. Davidson, *J. Am. Chem. Soc.,* **83,** 2599 (1961).

122. M. N. Williams and D. M. Crothers, *Biochemistry,* **14,** 1944 (1975).

123. (a) H. I. Heitner, S. J. Lippard, and H. R. Sunshine, *J. Am. Chem. Soc.,* **94,** 8936 (1972); (b) H. R. Sunshine and S. J. Lippard, *Nucleic Acids Res.,* **1,** 673 (1974).

124. B. C. Pal, L. R. Shugart, K. R. Isham, and M. P. Stulberg, *Arch. Biochem. Biophys.,* **150,** 86 (1972).

125. L. S. Lerman, *J. Mol. Biol.,* **3,** 18 (1961).

126. K. W. Jennette, Ph.D. Dissertation, Columbia University, 1975.

127. K. W. Jennette, S. J. Lippard, G. A. Vassiliades, and W. R. Bauer, *Proc. Natl. Acad. Sci. USA,* **71,** 3839 (1974).

128. M. Howe-Grant, K. C. Wu, W. R. Bauer, and S. J. Lippard, *Biochemistry,* **15,** 4339 (1976).

129. G. Scatchard, *Ann. N.Y. Acad. Sci.,* **51,** 660 (1949).

130. S. J. Lippard, P. J. Bond, K. C. Wu, and W. R. Bauer, *Science,* **194,** 726 (1976).

131. J. K. Barton and S. J. Lippard, *Biochemistry,* **18,** 2661 (1979).

132. M. Howe-Grant, Ph.D. Dissertation, Columbia University, 1978.

133. R. W. Davis, M. Simon, and N. Davidson, *Methods Enzymol.,* **21,** 413 (1971).

134. S. J. Singer, *Nature,* **183,** 1523 (1959).

135. S. K. Aggarwal, *J. Clin. Hematol. Oncol.,* **7,** 760 (1977), and references therein.

136. (a) M. Beer and C. R. Zobel, *J. Mol. Biol.,* **3,** 717 (1961); (b) M. Beer and E. N. Moudrianakis, *Proc. Natl. Acad. Sci. USA,* **48,** 409 (1962).

137. J. Wall, J. Langmore, M. Isaacson, and A. V. Crewe, *Proc. Natl. Acad. Sci. USA,* **71,** 1 (1974).

138. R. F. Whiting and F. P. Ottensmeyer, *J. Mol. Biol.,* **67,** 173 (1972).

139. R. K. Dale, E. Martin, D. C. Livingston, and D. C. Ward, *Biochemistry,* **14,** 2447 (1975).

140. R. F. Whiting and F. P. Ottensmeyer, *Biochim. Biophys. Acta,* **474,** 334 (1977).

141. C. H. Chang, M. Beer, and L. G. Marzilli, *Biochemistry,* **16,** 33 (1977), and references therein.

142. L. R. Subbaraman, J. Subbaraman, and E. J. Behrman, *Bioinorg. Chem.,* **1,** 35, (1971).

143. F. B. Daniel and E. J. Behrmann, *Biochemistry,* **15,** 565 (1976).

144. M. D. Cole, J. W. Wiggins, and M. Beer, *J. Mol. Biol.,* **117,** 387 (1977).

145. (a) L. G. Marzilli, B. E. Hanson, L. Kapili, S. D. Rose, and M. Beer, *Bioinorg. Chem.,* **8,** 531 (1978); (b) S. Rose, Middle Atlantic Regional Meeting, American Chemical Society, 1978.

146. (a) H. Matzura and F. Eckstein, *Eur. J. Biochem.,* **3,** 448 (1968); (b) F. Eckstein and K. H. Scheit, *Proc. Natl. Acad. Sci. USA,* **2,** 665 (1975).

147. K. G. Strothkamp and S. J. Lippard, *Proc. Natl. Acad. Sci. USA,* **73,** 2536 (1976).

148. J. D. Griffith, *Science,* **201,** 525 (1978).

149. F. Sanger and A. R. Coulson, *J. Mol. Biol.,* **94,** 441 (1975).

150. A. Maxam and W. Gilbert, *Proc. Natl. Acad. Sci. USA*, **74**, 560 (1977).

151. M. D. Cole, M. Beer, T. Koller, W. A. Strycharz, and M. Nomura, *Proc. Natl. Acad. Sci. USA*, **75**, 270 (1978).

152. K. G. Strothkamp, J. Lehmann, and S. J. Lippard, *Proc. Natl. Acad. Sci. USA*, **75**, 1181 (1978).

153. A. Jack, J. E. Ladner, D. Rhodes, R. S. Brown, and A. Klug, *J. Mol. Biol.*, **111**, 315 (1977).

154. P. J. Bond, R. Langridge, K. W. Jennette, and S. J. Lippard, *Proc. Natl. Acad. Sci. USA*, **72**, 4825 (1975).

155. W. Fuller and M. J. Waring, *Ber. Bunsenges, Phys. Chem.*, **68**, 805 (1964).

156. D. M. Crothers, *Biopolymers*, **6**, 575 (1968).

157. U. S. Nandi, J. C. Wang, and N. Davidson, *Biochemistry*, **4**, 1687 (1965).

158. N. Davidson, J. Wielholm, U. S. Nandi, R. Jensen, B. M. Olivera, and J. C. Wang, *Proc. Natl. Acad. Sci. USA*, **53**, 111 (1965).

159. P. J. Stone, A. D. Kelman, and F. M. Sinex, *Nature*, **251**, 736 (1974).

160. T. A. Beerman and J. Lebowitz, *J. Mol. Biol.*, **79**, 451 (1973).

161. R. M. K. Dale and D. C. Ward, *Biochemistry*, **14**, 2458 (1975).

162. M. M. Smith and R. C. C. Huang, *Proc. Natl. Acad. Sci. USA*, **73**, 775 (1976).

163. G. F. Crouse, E. J. Fodor, and P. Doty, *Proc. Natl. Acad. Sci. USA*, **70**, 1564 (1976).

164. T. J. C. Beebe and P. H. W. Butterworth, *Eur. J. Biochem.*, **66**, 543 (1976).

165. M. S. Kayne and M. Cohn, *Biochemistry*, **13**, 4159 (1974).

166. (a) J. M. Wolfson and D. R. Kearns, *Biochemistry*, **14**, 1436 (1975); (b) J. M. Wolfson and D. R. Kearns, *J. Am. Chem. Soc.*, **96**, 3653 (1974).

167. M. J. Cleare and J. D. Hoeschele, *Plat. Met. Rev.*, **17**, 1 (1973).

168. T. A. Connors, M. Jones, W. C. J. Ross, P. D. Braddock, A. R. Khokhar, and M. L. Tobe, *Chem.–Biol. Int.*, **5**, 415 (1972).

169. P. D. Braddock, T. A. Connors, M. Jones, A. R. Khokhar, D. H. Melzack, and M. L. Tobe, *Chem.–Biol. Int.*, **11**, 145 (1975).

170. F. R. Hartley, *The Chemistry of Platinum and Palladium*, Wiley, New York, 1973.

171. F. A. Cotton and G. Wilkinson, *Advanced Inorganic Chemistry*, Wiley, New York, 1972.

172. R. J. Speer, H. Ridgeway, D. P. Stewart, L. M. Hall, A. Zapata, and J. M. Hill, *J. Clin. Hematol. Oncol.*, **7**, 210 (1977).

173. M. C. Lim and R. B. Martin, *J. Inorg. Nucl. Chem.*, **38**, 1911 (1976).

174. J. A. Howle and G. R. Gale, *Biochem. Pharm.*, **19**, 2757 (1970).

175. H. C. Harder and B. Rosenberg, *Int. J. Cancer*, **6**, 207 (1970).

176. D. M. Taylor, J. D. Jones, and A. B. Robins, *Biochem. Pharm.*, **22**, 833 (1973).

177. J. Toth-Allen, Ph.D. Dissertation, Michigan State University, 1970.

178. S. Reslova, *Chem.–Biol. Int.*, **4**, 66 (1971).

179. B. Rosenberg, *J. Clin. Hematol. Oncol.*, **7**, 817 (1977).

180. D. J. Beck and R. R. Brubaker, *Mutation Res.*, **27**, 181 (1975).

181. P. Lecointre, J. P. Macquet, and L. L. Butour, Proc. Euchem Conference, Coordination Chemistry and Cancer Chemotherapy, Toulouse, 1978, Abstract 19; *Biochimie*, **60**, 1050 (1978).

182. D. J. Beck and R. R. Brubaker, *J. Bacteriol.*, **116**, 1245 (1972).

183. J. Drobnick, M. Urbankova, and A. Krakulova, *Mutation Res.*, **17**, 13 (1973).

184. H. W. van den Berg and J. J. Roberts, *Chem.–Biol. Intr.*, **11**, 493 (1975).

185. H. W. van den Berg and J. J. Roberts, *Mutation Res.*, **33**, 279 (1975).

186. J. J. Roberts, in *Scientific Foundation of Oncology*, T. Symington and R. L. Carter, Eds., Heinemann, London, 1976, p. 319.

187. J. M. Pascoe and J. J. Roberts, *Biochem. Pharm.*, **23**, 1345 (1974).

188. H. C. Harder and R. G. Smith, *Pharmacologist*, **16**, 253 (1974).

189. H. C. Harder, R. G. Smith, and A. F. Leroy, *Cancer Res.*, **36**, 3821 (1976).

190. N. P. Johnson, J. D. Hoeschele, N. B. Kuemmerle, W. E. Masker, and R. O. Rahn, *Nature*, in press.

191. L. L. Munchausen, *Proc. Natl. Acad. Sci. USA*, **71**, 4519 (1974).

192. J. J. Roberts and J. M. Pascoe, *Nature*, **235**, 282 (1972).

193. K. V. Shooter, R. Howse, R. K. Merrifield, and A. B. Robins, *Chem.–Biol. Int.*, **5**, 289 (1972).

194. N. P. Johnson, Middle Atlantic Regional Meeting, American Chemical Society, 1978.

195. R. Faggiani, B. Lippert, C. J. L. Lock, and B. Rosenberg, *J. Am. Chem. Soc.*, **99**, 777 (1977).

196. R. Faggiani, B. Lippert, C. J. L. Lock, and B. Rosenberg, *Inorg. Chem.*, **16**, 1191 (1977).

197. I. A. G. Roos, *Chem.–Biol. Int.*, **16**, 39 (1977).

198. J. L. Butour and J. P. Macquet, *Eur. J. Biochem.*, **78**, 455 (1977).

199. A. J. Thomson, R. J. P. Williams, and S. Reslova, *Structure and Bonding*, **11**, 1 (1972).

200. H. C. Harder, *Chem.–Biol. Int.*, **10**, 27 (1975).

201. K. V. Shooter and R. K. Merrifield, *Biochim. Biophys. Acta*, **287**, 16 (1972).

202. L. L. Munchausen and R. O. Rahn, *Biochim. Biophys. Acta*, **414**, 242 (1975).

203. G. L. Cohen, W. R. Bauer, J. K. Barton, and S. J. Lippard, *Science*, **203**, 1014 (1979).

204. J. P. Macquet and T. Theophanides, *Bioinorg. Chem.*, **5**, 59 (1975).

205. M. M. Millard, J. P. Macquet, and T. Theophanides, *Biochim. Biophys. Acta*, **402**, 166 (1975).

206. J. Dehand and J. Jordanov, *J.C.S. Chem. Commun.*, 598 (1976).

207. D. M. L. Goodgame, I. Jeeves, F. L. Phillips, and A. C. Skapski, *Biochim. Biophys. Acta*, **378**, 153 (1975).

208. G. P. Kuntz and G. Kotowycz, *Biochemistry*, **14**, 4144 (1975).

209. E. Sletten, *J.C.S. Chem. Commun.*, 558 (1971).

210. H. C. Harder and R. G. Smith, *J. Clin. Hematol. Oncol.*, **7**, 401 (1977).

211. I. A. G. Roos, A. J. Thomson, and S. Mansy, *J. Am. Chem. Soc.*, **96**, 6484 (1974).

212. P. J. Stone, A. D. Kelman, F. M. Sinex, M. M. Bhargava, and H. O. Halvorson, *J. Mol. Biol.*, **104**, 793 (1976).

213. A. D. Kelman, H. J. Peresie, M. Buchbinder, and M. J. Clark, Proc. Euchem Conference, Coordination Chemistry and Cancer Chemotherapy, Toulouse, 1978, Abstract 8; *Biochimie*, **60**, 1042 (1978).

214. B. K. Teo, P. Eisenberger, J. Reed, J. K. Barton, and S. J. Lippard, *J. Am. Chem. Soc.*, **100**, 3225 (1978).

215. J. K. Barton and S. J. Lippard, *Ann. N.Y. Acad. Sci.*, **313**, 686 (1978).

216. S. Wherland, E. Deutsh, J. Eliason, and P. B. Sigel, *Biochem. Biophys. Res. Commun.*, **54**, 662 (1973).

217. I. A. G. Roos, A. J. Thomson, and J. Eagles, *Chem.–Biol. Int.*, **8**, 421 (1974).

218. M. E. Friedman and J. E. Tiggins, *Fed. Proc.*, **35**, 623 (1976).

219. M. E. Friedman, P. Melius, J. E. Tiggins, and C. A. McAuliffe, *Bioinorg. Chem.*, **8**, 341 (1978).

CHAPTER **3**

Metals and
Genetic Miscoding

LAWRENCE A. LOEB and RICHARD A. ZAKOUR

The Joseph Gottstein Memorial Cancer Research Laboratory
Department of Pathology, University of Washington
Seattle, Washington

CONTENTS

1 INTRODUCTION

Certain metals have been shown to act as teratogens, mutagens, and carcinogens. Such deleterious effects by metals are of interest to biologists since metal ions are required by all living cells and metals function in DNA replication. The primary biological molecules directly involved in the transfer of genetic information are the DNA of the cell and the enzymes required for catalysis of the DNA replication process, particularly the DNA polymerases. All of the DNA polymerases that have been isolated and characterized to date require divalent metal ions for catalysis (1, 2). A number of different divalent metal ions can serve as polymerase activators in vitro (3). In general, toxicity to cells in the presence of high concentrations of metals has been attributed to the replacement of one metal with another so as to alter the catalytic functions of an enzyme. With regard to DNA polymerases, the fidelity of DNA replication in vitro is markedly affected by the type and amount of metal ions present during catalysis (3). In addition, current evidence suggests that the interactions between metal ions and the DNA template itself may also influence the accuracy of DNA replication (4, 5).

In this chapter we consider the role of metals in the transfer of genetic information. Since it is frequently easier to define normal cellular mechanisms by investigating pathologic or altered states, information about the accurate copying of genetic information from one generation to another can be derived by studying deficits in this accuracy which lead to mutations. Our approach to this problem is to examine DNA synthesis in vitro, to determine the effects of different metal ions on the fidelity of this synthesis, and then to ask whether alterations in the fidelity of DNA synthesis are related to the mutagenic and carcinogenic properties of metals.

2 RELATIONSHIP BETWEEN MUTAGENESIS AND CARCINOGENESIS

Boveri (6) postulated that somatic mutations provide the basis for carcinogenesis. Evidence to support this hypothesis is as follows: (a) the target of most carcinogens is DNA (7); (b) most chemical carcinogens, when activated to reactive electrophiles, are mutagenic (8–10); (c) malignant changes are frequently associated with chromosomal abnormalities (11); (d) the malignant phenotype is in nearly all cases permanent and heritable in cells (12); (e) certain human diseases with defects in DNA repair have an unusually high incidence of malignancy (13, 14); and (f) certain inherited diseases are associated with malignancy (15). Most of the above evidence can also be used to support an alternative hypothesis, namely, that alterations in the expression of genes in the absence of mutations are the underlying basis for malignancy. However, the observations that most carcinogens can be identified as mutagens provide strong support for the concept of somatic mutations as the basis of carcinogenesis. Practically, the rapid technology for analyzing mutagenesis in vitro and for cloning individual genes has provided the experimental tools for testing the somatic mutation hypothesis.

Although a causal relationship between mutagenesis and carcinogenesis has not been firmly documented, the mechanisms of each of these processes are currently being investigated with respect to metals. It is therefore appropriate to include a summary of the studies concerning metal mutagenesis and carcinogenesis. The postulated mechanisms of metal involvement in these processes is discussed later in this chapter.

3 METALS AS MUTAGENS

Metals have been shown to be mutagenic in both procaryotic and eucaryotic organisms. The initial efforts to demonstrate the mutagenicity of metal ions in bacterial systems were inconclusive. Presumably this was due to the large volume of culture medium relative to bacterial cell mass and to the presence of competing metal ions in the media. More recently, however, positive results have been obtained. Evidence for the mutagenic characteristics of certain metal compounds has been summarized in recent reviews by Flessel (16) and Sunderman (17). Compounds containing arsenic, chromium, copper, iron, manganese, molybdenum, platinum, and selenium ions are mutagenic in many bacteria (Table 1). Additonally, compounds of arsenic, chromium, and molybdenum diminish DNA repair in recombination-deficient strains of *Bacillus subtilis* (18). Also, deleterious effects in eucaryotic cells have been associated with compounds of aluminum, arsenic, antimony, beryllium, cadmium, cobalt, chromium, iron, lead, mercury, nickel, and tellurium (Table 2).

The nature of the mutational changes induced by metal ions in bacteria has been surmised from studies with different tester strains. For example, chromium, manganese, and iron compounds revert frameshift mutations (16, 19, 23). Manganese sulfate also produces point mutations (34). Selenium (20) and platinum (24) compounds cause base pair substitutions. Although not yet demonstrated, it is also possible that some metal mutagenesis may be the result of suppressor mutations. Outside of evi-

Table 1   Metal Mutagenesis in Bacteria

Metal	Organism	Selected References
As	*B. subtilis*	18
Cr	*B. subtilis*	18
	E. coli	19
	S. typhimurium	20
Cu	*B. subtilis*	21
Fe	*E. coli*	22
	S. typhimurium	23
Mn	*S. typhimurium*	16
Mo	*B. subtilis*	18
Pt	*S. typhimurium*	24
Se	*S. typhimurium*	20

dence for increased GC to AT transition caused by chromate ions in *Escherichia coli* (19), little is known about the base changes that occur during metal mutagenesis. Furthermore, the effects of nearest-neighbor nucleotide sequence and DNA structure remain to be investigated in relationship to mutagenesis.

Table 2   Metal-Induced Damage in Eucaryotic Cells[a]

Metal	Chromosomal Aberration	Mitotic or Cell Division Disturbance	Morphologic Alteration
Al	25		
As	26		27
Be			27
Cd	25, 28		27
Co		29	
Cr	30		27, 31
Fe		29	
Hg		32	
Ni		29	27
Pb	33		27
Sb	26		
Te	26		

[a]Selected references.

4 METALS AS CARCINOGENS

4.1 Metal Carcinogenesis in Mammals

A number of studies have indicated that metals implanted in various tissues in different experimental animals result in tumorigenesis (Table 3). The tumors induced by metal implantation are malignant in that they invade, metastasize, and ultimately kill the host animal. Of the metals studied, nickel has been most extensively investigated, in particular by Sunderman and colleagues. Sarcomas have been produced in rodents after nickel implantation subcutaneously, intramuscularly, intrafemorally, and intrathorasically (17). In addition, inhalation of nickel carbonyl gas resulted in the production of squamous cell carcinomas in the lungs of rodents (44) and caused teratogenic effects on the progeny of pregnant rats (46).

The induction of tumors by metals was initially considered a curiosity, an example of "solid state" carcinogenesis. Since a number of inert

Table 3 Metal Carcinogenesis in Experimental Animals

Metal	Route[a]	Species	Resultant Tumor	Selected References
Beryllium	Inhalation	Rat	Pulmonary carcinoma	35
	IV	Monkey	Osteosarcoma	36
Cadmium	IM	Rat	Sarcoma	37
Chromium	SC	Rabbit	Liposarcoma	38
Cobalt	SC	Rat	Sarcoma	37
Iron[b]	IM, IP, SC	Rat	Sarcoma	40
Lead	PO	Mouse	Renal carcinoma	41
Manganese	IP	Mouse	Pulmonary adenoma	42
Nickel	IM	Rat	Sarcoma	43
	Inhalation	Pig	Pulmonary carcinoma	44
Titanium[c]	IM	Rat	Lymphoma	45
Zinc[d]	IT	Hamster	Teratoma	37

[a]IM = Intramuscular; IP = intraperitoneal; IT = intratesticular; IV = intravenous; PO = oral; SC = subcutaneous.
[b]In this report only iron carbohydrate complexes induced sarcoma.
[c]In this report only an organic titanium compound at very high concentrations induced lymphoma.
[d]Carcinogenesis resulted only when zinc salts were injected directly into the gonads.

plastics have been shown to produce tumors at the site of implantation, it was argued that the simple distortion of cell geometry by any inert foreign substance might lead to the production of tumors. It is unlikely, however, that "solid state" carcinogenesis is the mechanism by which metals induce tumors since the production of tumors by metals has been demonstrated to be independent of the physical state or geometry of the implanted metal. Furthermore, tumors have been shown to result from the injections of ionic nickel compounds (47).

It is difficult to identify the ionic species associated with the induction of malignancy. Presumably metals are oxidized in cells to the cations that initiate tumorigenesis. Thus considering the complexity of cells, the multiple ionic states of metals, and the number of metal molecules that may be required for tumorigenesis, it is a reasonable expectation that the relevant ionic species of metals involved in carcinogenesis will be identified only in simple in vitro systems.

4.2 Metals and Human Cancer

Epidemiologic evidence indicates that As, Cr, and Ni and possibly Be, Cd, Fe, and Pb are human carcinogens (Table 4). The respiratory cancer risk in workers exposed to arsenic compounds is as great as 12 times that in control populations (54). In studies in both the United States and the USSR, the incidence of respiratory cancer among chromate refinery workers, chromate pigment workers, and chrome platers is 10 times greater than in control groups (55, 17). Studies on Ni are most extensive. The incidence of lung cancer and cancer of the nasal sinus in nickel refinery workers prior to automation is reported to be five and 243 times that in the general population (52, 53). Also, many of these trace metals are abundant in asbestos (56) and might account for its carcinogenicity.

Table 4 Metal Carcinogenesis in Humans

Metal	Exposure	Site of Tumor	Selected References
Arsenic	Agricultural	Gastrointestinal, lung tumor	48
	Metallurgical	Lung	49
	Chemical	Lung	50
Chromium	Painters	Lung	51
Nickel	Refinery	Pulmonary	52
		Nasal sinus	53

5 RELATIONSHIP BETWEEN METAL MUTAGENESIS AND CARCINOGENESIS

As stated previously, most agents that initiate malignancy have been shown to be mutagenic. Some of the metals that are unequivocal carcinogens (As, Be, Cd, Co, Cr, Mn, Ni, Pb) either in experimental animals or in humans have been assayed and are mutagenic in bacteria (As, Cr, Mn) or produce chromosomal abnormalities in eucaryotes (As, Cd, Cr, Ni, Pb). Thus metal carcinogens are no exception to the general postulate that carcinogens can be detected by their ability to cause mutations.

In general, the relationship between somatic mutation and carcinogenesis is not well understood. In Figure 1 we present several possibilities for the progression of metal mutagenesis to carcinogenesis. Metal-induced mutations may occur by the interaction of metal ions with the DNA template or with the DNA polymerase. In the latter case a normal polymerase could be exposed to an abnormal concentration of physiologically required metals or to exogenous metals that are usually not present during cellular metabolism. Alternatively, metal ions that are not normally used for DNA replication could serve as activators for a DNA polymerase that has been previously altered. In either case an abnormal polymerase–metal combination might decrease the fidelity with which the DNA is replicated and thus lead to the synthesis of DNA that contains mutations. This newly synthesized DNA may contain certain

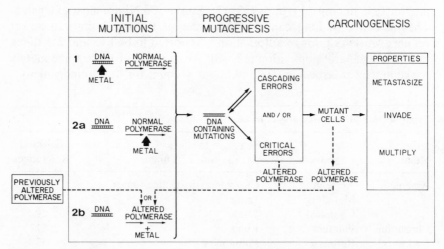

Figure 1 Speculative models for the involvement of metals in mutagenesis and carcinogenesis. Metal ions may interact with (1) the DNA template or (2) the DNA polymerase (see text).

critical errors (e.g., genes that code for altered polymerases). Furthermore, continued replication of the DNA by an altered polymerase or in the presence of mutagenic metals could also lead to an accumulation of additional errors during subsequent rounds of replication. Such critical errors and/or cascading errors caused by an accumulation of mutations may account for the progressive change in cellular properties during tumor progression (7, 58).

6 THE EFFECT OF METALS ON IN VITRO DNA SYNTHESIS

The phenomenologic correlations we have mentioned suggest that exposure of cells to certain metal ions diminishes the accuracy of DNA replication. In vivo systems are too complicated to begin to unravel the mechanism by which metal ions diminish the fidelity of DNA synthesis and the mechanisms by which metal ions induce mutations. Our approach to this problem has been to examine DNA synthesis in vitro, to determine the effects of different metal ions on the fidelity of this synthesis, and then to ask whether alterations in the fidelity of DNA synthesis are related to the mutagenic and carcinogenic properties of these metals.

Prior to considering these studies in detail it is instructive to consider what are the requirements for the fidelity of DNA synthesis in cells, the role of metal ions in the mechanism of DNA polymerization in vitro, and methods for measuring fidelity of DNA synthesis.

7 FIDELITY OF DNA SYNTHESIS

In vitro model systems for assaying fidelity provide a view of an exceedingly accurate biological mechanism. On the basis of spontaneous mutation rates in procaryotic and eucaryotic cells, stable misincorporation of a base during DNA replication is estimated to occur with a frequency of 10^{-8} to 10^{-11} per base pairs synthesized (59). This accuracy appears to be achieved by a multistep process (Table 5) (60). The differences between correct and incorrect Watson–Crick base pairings involve only one or two hydrogen bonds. This difference in free energy, ΔG, has been estimated to account for an error rate of approximately 10^{-2} (61, 62, 58). (The effects of metallic ions on base pairings per se—in the absence of polymerization—is considered in another chapter of this book.) DNA polymerases also participate in base selection, reducing the error rate to values approaching 10^{-5} (2). Two different categories of mechanisms have been considered to account for the role of DNA polymerases in enhancing

Table 5   Contributions to the Fidelity of DNA Replication

Step	Error Rates
1. Base pairing	10^{-2}
2. Fidelity of purified DNA polymerases	10^{-3} to 10^{-5}
A. Error correction (procaryotic polymerases)	
B. Error prevention (reverse transcriptase and eucaryotic DNA polymerases)	

	Mutation Rate
3. Fidelity in vivo	10^{-8} to 10^{-11}

Possible Additional Selective Factors

A. Enhanced error prevention
   1. DNA binding proteins
   2. Polymerase binding proteins
B. Repetitive error correction
   1. Exonucleases in eucaryotic cells
C. Postreplicative excision of mismatched base pairs
   1. Specific endonucleases

base selection: (1) Procaryotic DNA polymerases contain an exonuclease that is able to excise mismatched bases at or immediately after incorporation (63). Such an exonucleolytic activity provides a means by which polymerases can "proofread" for the insertion of incorrect bases during catalysis. (2) Eucaryotic DNA polymerases lack such an exonuclease activity (64). Thus enhanced base selection by these enzymes must represent increased base selectivity at the catalytic site. The effect of metal ions on the accuracy of DNA synthesis at the level of DNA synthesis in vitro is extensively considered in this chapter.

In order to increase the accuracy achieved by DNA polymerases in vitro (10^{-3} to 10^{-5}) to the level that occurs during DNA replication in vivo (10^{-8} to 10^{-11}), additional base-selective factors must also play a role. Based on the method of catalysis by purified DNA polymerases and their known interactions with other cellular proteins, three hypothetical categories have been proposed to account for the increased accuracy:

1. In accord with the concept of error prevention, correct nucleotide selectivity may be achieved by factors that increase the rigidity of the substrate–template conformation. Candidates for such factors are possibly DNA-binding proteins and/or proteins that bind to DNA polymerases.

2. By a repetitive proofreading exonuclease, any level of fidelity

theoretically can be achieved. In animal cells, it is possible that exonucleases not part of the polymerase molecule serve such a function.

3. Excision of mismatched bases by specific endonucleases could greatly enhance fidelity. In this case the repaired segment would have to be sufficiently small so as not to allow additional mistakes during synthesis of DNA by the polymerase.

The effects of metal ions on the fidelity of DNA synthesis can be evaluated at the level of polynucleotides by measurements of nucleotide interactions, at the level of DNA synthesis by quantitation of incorrect nucleotide incorporation, and only retrospectively at the level of DNA replication by measuring mutation frequencies.

8   MECHANISM OF DNA POLYMERIZATION

Generically, DNA polymerases have similar requirements for catalysis (Fig. 2). Synthesis proceeds by a sequential addition of nucleotide monomers (deoxynucleoside monophosphates) with a concomitant release of pyrophosphate (1). DNA polymerases are part of a unique class of enzymes in that they primarily take direction from another molecule, a template (Fig. 2). In cells, the template is DNA. Only in the case of DNA polymerases from RNA tumor viruses (reverse transcriptase) has RNA

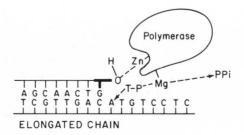

Figure 2   DNA polymerase reaction (65).

been unambiguously demonstrated to serve as a template for DNA synthesis in vivo. Synthetic polydeoxynucleotides and polyribonucleotides can serve as templates in vitro with most DNA polymerases. Two DNA polymerases, reverse transcriptase (66) and *E. coli* DNA polymerase I (67), are able to copy natural RNA templates in vitro. In all cases synthesis is started on the 3'-hydroxy terminus of the primer strand hybridized onto the template strand. The primer can be an oligonucleotide, one strand of double-stranded DNA, or a hairpin loop of single-stranded DNA. Thus DNA polymerases only elongate already existing polynucleotide chains; they cannot initiate chains de novo as do RNA polymerases. The substrates of all known DNA polymerases are deoxynucleoside triphosphates that are complementary to the template. Based on the similar requirements for activity and a spectrum of similar kinetic parameters, it is a reasonable expectation that there is a common mechanism for catalysis by DNA polymerases from different sources (2). The lower portion of Figure 2 is one of the current models for the detailed coordination of the enzyme, primer, and nucleotide substrate (65).

The added divalent ion Mg^{2+} has been shown to coordinate the enzyme with the substrate in the form of an enzyme–metal–substrate complex (68). Mn^{2+}, Co^{2+}, Ni^{2+}, and in certain cases Zn^{2+} have been shown to substitute for Mg^{2+} as activators (3). Analysis of *E. coli* DNA polymerase I–Mn^{2+}–substrate complexes indicates that the enzyme in the absence of template alters the conformation of the deoxynucleoside triphosphate substrate to that which it would occupy in double-helical DNA (68). The function of the metal activator in the DNA polymerase reaction per se has recently been extensively reviewed (60, 61).

Evidence also suggests that DNA polymerases are zinc metalloenzymes. *Escherichia coli* DNA polymerase I has been shown to contain 1 gram-atom of zinc per mole (69). Removal of the zinc by chelation is accompanied by a proportional loss of DNA polymerase activity. Restoration of the activity occurs upon the readdition of zinc. Similar but less stringent criteria indicate that DNA polymerase from avian myeloblastosis virus (AMV) is also a zinc metalloenzyme (70). The presence of zinc has been demonstrated in a variety of homogeneous DNA polymerases. Furthermore, a number of DNA polymerases have been shown to be inhibited by chelators such as *o*-phenanthroline but not its nonchelating analog *m*-phenanthroline (60, 61). This latter criterion has been questioned by the studies of Sigman and co-workers (71), who found that the inhibition by *o*-phenanthroline was due to the formation of a complex with copper, which was a contaminant in the thiol compounds present in the assay. Presently there is no evidence to suggest that interactions with enzyme-bound zinc alter fidelity (72).

9 ASSAYS OF FIDELITY WITH POLYNUCLEOTIDE TEMPLATES

Until very recently all assays of the fidelity of DNA synthesis in vitro measured the ability of DNA polymerases to copy homopolymer or alternating copolymer templates. These templates contained only one or two nucleotides, and the mismatched nucleotide was identified simply as one not complementary to the template nucleotides. Using this assay one can observe the effects of both activating and nonactivating metals on the fidelity of DNA synthesis. As a template, for critical measurements of fidelity, we have chosen poly[d(A–T)], a synthetic polynucleoside consisting of deoxythymidine and deoxyadenosine monophosphates (Fig. 3) (73, 74). Poly[d(A–T)] can be synthesized to contain less than one in 2 ×

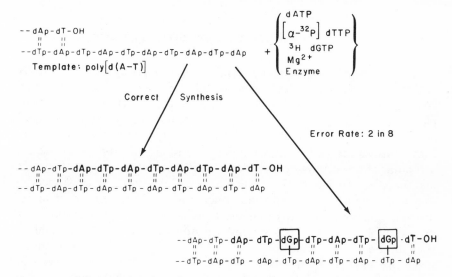

Figure 3 Fidelity assay using synthetic polynucleotide templates. Reaction mixture for a typical fidelity assay using poly[d(A–T)] as a template and dGTP as the noncomplementary nucleotide is as follows: Assays are carried out in a total volume of 0.05 ml containing the following: 100 mM Tris-HCl, pH 7.5; 60 mM KCl; 1.0 mM MgCl₂; 20 μM [α ³²P]-dTTP (13 dpm/pmol); 20 μM dATP; 20 μM [³H]-dGTP or [³H]-dCTP, 0.2 mM poly[d(A–T)]; and sufficient purified DNA polymerase to incorporate 300 pmol of the complementary nucleotide. All reactions are performed in triplicate and incubated at 37°C for 15 min. Incorporation of the radioactive deoxynucleoside triphosphates into an acid-insoluble product is determined after repeatedly precipitating the polynucleotide product with 1.0 N perchloric acid and 50 mM sodium pyrophosphate and solubilizing with 0.2 M NaOH at 20°C as previously described (73). The final precipitate is collected onto glass fiber discs, and radioactivity is measured by scintillation spectroscopy. After subtracting the amount of radioactivity in nonincubated controls, the error frequency is calculated from the ratio of the noncomplementary to the total complementary nucleotides incorporated.

10^6 mistakes using a de novo reaction with *E. coli* DNA polymerase I (74). Copied correctly, only dAMP and dTMP would be incorporated into the newly synthesized product. By using α^{32}P-dTTP, unlabeled dATP, and ^{3}H-dGTP or ^{3}H-dCTP, one can simultaneously measure the incorporation of complementary and noncomplementary nucleotides (73). The incorporation of either dCTP or dGTP would represent errors. The frequency of misincorporation is obtained from the ratio of ^{3}H to ^{32}P in the acid-insoluble product. Control experiments are required to show that the tritium label in the reaction product is in the noncomplementary nucleotides and not in any radioactive contaminants. Also, it must be demonstrated that the noncomplementary nucleotides are covalently incorporated in phosphodiester linkage. Using nearest-neighbor analysis one can determine the distribution of the noncomplementary nucleotides.

10   FIDELITY OF DNA POLYMERASES WITH POLYNUCLEOTIDES

Measurements of the frequency of misincorporation by DNA and RNA polymerases when copying polynucleotide templates carried out in this laboratory (75) are listed in Table 6. In these experiments the complementary and noncomplementary nucleotides were present at equal concentrations corresponding to the K_m of the complementary nucleotide. Each assay included activating concentrations of Mg^{2+} and nearly saturating concentrations of template. In all cases incorporation of correct and incorrect nucleotides was shown to be linear with time of incubation and proportional to enzyme concentration. Furthermore, nearest-neighbor analysis indicated that the noncomplementary nucleotide was invariably incorporated as a single-base substitution. The most frequent substitution was a purine for a purine or a pyrimidine for a pyrimidine. From this tabulation one observes that different DNA polymerases can copy the same templates with differing fidelity. Perhaps the most striking feature of this analysis is that the error rates of the procaryotic DNA polymerases, those with a $3' \rightarrow 5'$ exonuclease, are similar to those of the polymerases that do not have an accompanying exonuclease. Thus the exonuclease in procaryotic DNA polymerases is not necessarily the major determinant of fidelity.

In general, DNA polymerase β from a variety of sources is more accurate than DNA polymerase α. Interestingly, circumstantial evidence suggests that DNA polymerase -α participates in DNA replication and that DNA polymerase -β functions in DNA repair (76). The high error rates of DNA polymerases from RNA tumor viruses has facilitated studies on the effects of metals on the incorporation of noncomplementary

Table 6 Fidelity of DNA Polymerases in Copying Polynucleotide Templates

DNA Polymerase	Template	Noncomplementary Nucleotide	Error Rate
Procaryotic DNA polymerase			
E. coli polymerase I	poly[d(A–T)]	dGTP	1/70,000
Bacteriophage T₄	poly[d(A–T)]	dCTP	1/12,000
Eucaryotic DNA polymerase			
Sea Urchin nuclei -α	poly[d(A–T)]	dGTP	1/12,000
Human placenta -α	poly[d(A–T)]	dGTP	1/9,000
Human lymphocyte -α	poly[d(A–T)]	dGTP	1/14,000
Calf thymus -α	poly[d(A–T)]	dGTP	1/9,000
Human placenta -β	poly[d(A–T)]	dGTP	1/40,000
Calf thymus -β	poly[d(A–T)]	dGTP	1/30,000
RNA polymerase			
E. coli	poly[d(A–T)]	dCTP	1/2,400
E. coli	poly[d(A–T)]	dGTP	1/42,000
Reverse transcriptase			
Avian myeloblastosis virus	poly(A)·oligo(dT)	dCTP	1/300–800
Avian myeloblastosis virus	poly[d(A–T)]	dGTP	1/3,000
Rous sarcoma virus	poly(C)·oligo(dG)	dATP	1/900

nucleotides. Because of the high frequency of mistakes, it has been possible to accurately measure the distribution of noncomplementary nucleotides in the reaction product.

11   EFFECTS OF ACTIVATING METAL IONS ON FIDELITY

Mn^{2+}, Co^{2+}, and Ni^{2+}, in addition to Mg^{2+}, serve as metal activators for DNA polymerases from animal, viral, and bacterial sources (3). Minimal activity has also been reported with Zn^{2+}. The ability of these metals to serve as activators of AMV DNA polymerase, with activated DNA as a template, is shown in Figure 4. The maximal rate of nucleotide incorporation with Mn^{2+}, Co^{2+}, and Ni^{2+} was 65%, 25%, and 7%, respectively, of that achieved with Mg^{2+} (77). Similar results have been obtained with *E. coli* DNA polymerase I (78).

The effects of Mg^{2+} and Mn^{2+} concentrations on the incorporation of complementary and noncomplementary nucleotides are illustrated in Figure 5 with AMV DNA polymerase and poly(rC) as a template (79). At activating concentrations of Mg^{2+} (6 mM), AMV DNA polymerase incorporates 1 molecule of dAMP for every 1400 molecules of dGMP polymerized. This error rate is invariant with respect to Mg^{2+} concentration. With Mn^{2+} at activating concentration (1 mM), the error rate was greater (1 in 800). At greater than activating concentrations of Mn^{2+}, there

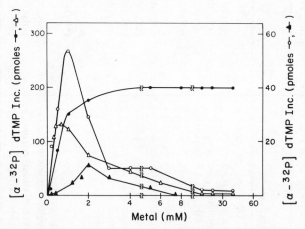

Figure 4   Metal activation of DNA synthesis with avian myeloblastosis virus DNA polymerase. Activity was measured using activated calf thymus DNA as a template and the following metal ions added as chloride salts: mg^{2+} (●), Co^{2+} (○), Mn^{2+} (△), and Ni^{2+} (▲). Results taken from Sirover and Loeb (78), with permission.

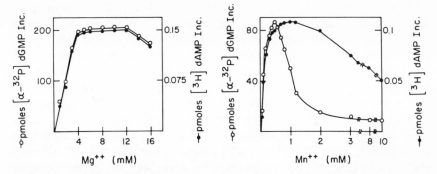

Figure 5 Effect of Mg^{2+} and Mn^{2+} on the fidelity of DNA synthesis. With poly (rC) · oligo (dG) as a template primer for AMV DNA polymerase, fidelity assays were carried out using [α ^{32}P]-dGTP as the complementary nucleotide and [^{3}H]-dAMP as the non-complementary nucleotide (79).

was a progressive decrease in the incorporation of the complementary nucleotide but not of the noncomplementary nucleotide, thus yielding a further increase in the frequency of misincorporation. At concentrations as great as 5 mM, the error rate approached 1 in 30, and nearest-neighbor analysis indicated that each misincorporation occurred as a single-base substitution. The decreased fidelity with increased Mn^{2+} concentration has been observed with all templates and noncomplementary nucleotides tested as well as with noncomplementary nucleotide substrates. An absolute increase in the rate of incorporation of the noncomplementary nucleotide can be demonstrated by simply using more DNA polymerase in the assay or prolonging the time of incubation. These data indicate that Mn^{2+} enhances infidelity with AMV DNA polymerase.

The error rates observed with Mg^{2+}, Mn^{2+}, and Co^{2+} using different DNA polymerases are shown in Table 7. These assays were carried out

Table 7 Effect of Metal Activator on Fidelity

DNA Polymerase	Reference	Mg^{2+} 5 mM	Mn^{2+} 0.1 mM	Mn^{2+} 2 mM	Co^{2+} 0.4 mM	Co^{2+} 5 mM
AMV	78	1/1,680	1/760	1/500	1/1,100	1/200
E. coli I	77	1/20,000	1/10,000	1/1,000	1/7,500	1/7,000
Human placenta -α	80	1/6,000	1/1,900	1/300	1/1,300	1/450
Human placenta -β	80	1/20,000	1/9,000	1/2,000	1/5,000	1/1,300

with alternating poly[d(A–T)] as a template and with dGTP as the non-complementary nucleotide. The increase in infidelity with Mn^{2+} using AMV DNA polymerase, *E. coli* DNA polymerase I, and DNA polymerases -α and -β is in agreement with the results of other investigations and may reflect a common aspect to the mechanism of fidelity by DNA polymerases. Substitution of Mn^{2+} for Mg^{2+} results in an increase in misincorporation by *E. coli* DNA polymerase I (81), T_4 DNA polymerase (82), and DNA polymerase -α (83, 80). The fact that Mn^{2+} and Co^{2+} alter the fidelity of the DNA polymerases that do not have an associated exonuclease [AMV (84), DNA polymerases -α and -β (64, 2)] indicates that these metal ions do not promote misincorporation by inhibiting an error correcting exonucleolytic activity. Ni^{2+} can also substitute for Mg^{2+} as a metal activator. However, the amount of synthesis achieved with Ni^{2+} as the metal activator has not been sufficient to accurately measure the changes in the fidelity of DNA synthesis with any DNA polymerase except *E. coli* DNA polymerase I, in which case Ni^{2+} promotes misincorporation (78).

In order to relate the measurements with alternate metal activators to a situation that would be expected to occur in cells, the effects of these activators on Mg^{2+}-activated DNA synthesis have been investigated. Co^{2+}, Mn^{2+}, and Ni^{2+} have been shown to enhance misincorporation by AMV DNA polymerase (77) and *E. coli* DNA polymerase I (78) in the presence of activating amounts of Mg^{2+}. Thus these metal activators could alter the fidelity of DNA polymerases in cells. In contrast to DNA polymerases, the substitution of Mn^{2+} for Mg^{2+} increases the accuracy of RNA synthesis with *E. coli* RNA polymerase (85). Additional studies concerning the effects of metal activators on fidelity with RNA polymerases are required before any mechanistic differences can be evaluated.

12   EFFECTS OF NONACTIVATING METAL CATIONS ON THE FIDELITY OF DNA SYNTHESIS

Beryllium, a known carcinogen, has been shown to decrease the fidelity of catalysis with *Micrococcus luteus* DNA polymerase (86) and AMV DMA polymerase (87). Be^{2+} is unable to substitute for Mg^{2+} as a metal activator. However, as a nonactivating cation, Be^{2+} alters the fidelity of DNA synthesis in the presence of Mg^{2+}. Preincubation of the enzyme but not the template, primer, or substrates with high concentrations of Be^{2+} resulted in an increased error rate. This finding suggests that Be^{2+} can interact with some noncatalytic site on DNA polymerase and thereby alter the fidelity of DNA synthesis. Be^{2+} has been shown to alter the fidelity of DNA polymerase -α from human fibroblasts (88), DNA polymerases -α and -β from human placenta, and $E.$ $coli$ DNA polymerase I (80). Since the eucaryotic and viral DNA polymerases lack an exonuclease, these results mitigate against the possibility that Be^{2+} interacts with the exonucleolytic site on procaryotic DNA polymerases.

13   INFIDELITY OF DNA SYNTHESIS: A BIOCHEMICAL SCREENING SYSTEM FOR METAL MUTAGENS AND CARCINOGENS

The experimental results that have been considered on the alteration in the fidelity of DNA synthesis by activating and nonactivating divalent metal cations and the dynamic nature of metal–macromolecular interactions prompted Sirover and Loeb (89) to ask if mutagenic and/or carcinogenic metals could be identified by alterations in the fidelity of DNA synthesis. So far, about 40 metal compounds have been tested in graded concentrations in this cell-free system. The method of analysis and the results are summarized in Figure 6. Twenty-two of these metal salts were

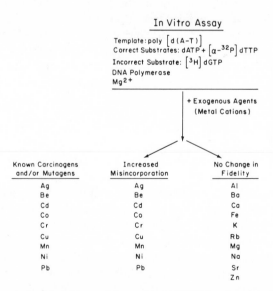

FIDELITY ASSAY
SCREEN FOR MUTAGENS AND/OR CARCINOGENS

In Vitro Assay

Template: poly $\left[d(A-T) \right]$
Correct Substrates: dATP + $\left[\alpha^{-32}P \right]$ dTTP
Incorrect Substrate: $\left[^{3}H \right]$ dGTP
DNA Polymerase
Mg^{2+}

+ Exogenous Agents
(Metal Cations)

Known Carcinogens and/or Mutagens	Increased Misincorporation	No Change in Fidelity
Ag	Ag	Al
Be	Be	Ba
Cd	Cd	Ca
Co	Co	Fe
Cr	Cr	K
Cu	Cu	Rb
Mn	Mn	Mg
Ni	Ni	Na
Pb	Pb	Sr
		Zn

Figure 6 Summary of results with metal ions using AMV DNA polymerase. Increased misincorporation represents a change of greater than 30% at two or more concentrations of the indicated metal cation.

tested in a triple-blind study in which the assays, computations, and designation of each unknown compound with respect to fidelity were carried out independently (89). Compounds which increased infidelity by greater than 30% at two or more concentrations were scored as positive. Metals were designated as carcinogens or mutagens by an evaluation of the literature before assessment of their effects on fidelity. An enhancement in the infidelity of DNA synthesis was observed with all the known mutagens and/or carcinogens tested (Ag, Be, Cd, Co, Cr, Mn, Ni, and Pb). The evidence in the literature on the mutagenicity or carcinogenicity of three of the metal cations was considered equivocal. Of these, Cu^{2+} increased misincorporation; Fe^{2+} and Zn^{2+} did not alter fidelity. All other metal salts tested were considered to be neither carcinogenic nor mutagenic, and they did not increase misincorporation. Only a few of the metal salts that did not alter fidelity are listed in Figure 6. If one considers the metals that are questionable as possible mutagens and carcinogens (Cu, Fe, Zn) to have a biological effect that is the opposite of that observed in the fidelity assay, the worst possible situation, namely, the hypothesis that increases in fidelity and mutagenicity/carcinogenicity are independent variables, can be rejected with a $p = 1.6 \times 10^{-4}$.

With only a few exceptions these results have been confirmed by Miyaki et al. (90) and Sirover et al. (78) using $E.$ $coli$ DNA polymerase I, and by Seal et al. using DNA polymerases -α and -β from human placenta (80). Pb^{2+}, Cd^{2+}, Co^{2+}, Cu^{2+}, and Mn^{2+} have been shown to stimulate chain initiation by RNA polymerases (91), whereas Zn^{2+}, Mg^{2+}, Li^+, Na^+, and K^+ were inhibitory. The similarity between the effects caused by particular metal ions on fidelity with DNA polymerases and on chain initiation with RNA polymerase could point to metal interactions with the DNA template as a common underlying mechanism (vida infra).

14   FIDELITY OF DNA SYNTHESIS WITH NATURAL DNA TEMPLATES

All of the aforementioned studies on the fidelity of DNA synthesis depended on measuring either incorporation of noncomplementary nucleotides using synthetic polynucleotide templates of limited composition or measuring the incorporation of nucleotide analogs. It has been assumed that the results with these model systems are similar to those that would be obtained copying natural DNA containing all four bases. It is known that during synthesis of template primers, with repeating nucleotide sequence, there is slippage of the primer relative to the template (1, 92). Thus metal-mediated changes in the fidelity of DNA synthesis could result from altered slippage, an event that presumably does not occur during

MUTATION RATE DURING DNA SYNTHESIS

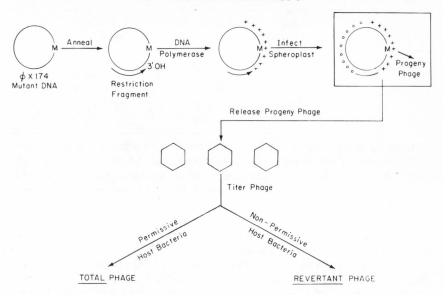

Figure 7 Fidelity of DNA synthesis with a natural DNA template. The mutation rate (i.e., reversion frequency of a previously existing mutation) is the proportion of revertant progeny phage among total progeny phage population. Alterations in the fidelity of DNA replication are detected by changes in the mutation rate.

copying of natural DNA templates. Also unique to homopolymers or repeating heteropolymers is the fact that a single noncomplementary nucleotide can occupy a looped-out structure without changing the reading frame for subsequent codons. Thus metals could enhance misincorporation by increasing the frequency of such looped-out structures. To circumvent these limitations, a system has been recently developed (93) to monitor the fidelity of in vitro DNA synthesis using a natural DNA template from the bacteriophage ϕX174 carrying a suppressible nonsense mutation, amber 3 (*am*3) (Fig. 7). Certain nucleotide substitutions within the *am*3 locus during in vitro replication of this DNA will cause a reversion to the wild-type phenotype. Transfection with the in vitro replicated ϕX174 DNA of *E. coli* spheroplasts under nonsuppressive conditions permits one to assay for progeny phage revertant for the *am*3 mutation. Thus measurement of the reversion frequency of the progeny phage indicates the accuracy with which the DNA in the region of this mutation was copied.

Preliminary estimates on the error rate of DNA polymerases in copying natural DNA templates in vitro are shown in Table 8 (94). With homoge-

Table 8 Fidelity of DNA Polymerases Copying φX174 DNA[a]

Polymerase	Metal Activator	Nucleotides per Template	Reversion Frequency	Error Rate
Control	Mg^{2+}	0	2.41×10^{-5}	—
E. coli Pol I	Mg^{2+}	610	3.35×10^{-5}	1/13,800
E. coli Pol I	Mn^{2+}	962	11.7×10^{-5}	1/1,390
E. coli Pol I	Co^{2+}	1350	5.99×10^{-5}	1/3,730
AMV	Mg^{2+}	500	20.4×10^{-5}	1/590

[a]The indicated DNA polymerases and metal activators were used to copy 6.4×10^{10} molecules of φX174 am3 DNA hybridized to the Z-5 Hae3 restriction fragment. The control DNA was subjected to all of the conditions of the reaction except that the DNA polymerase was omitted. Each reaction contained 30 μM of all four deoxynucleoside triphosphates. The metal activators were present as chloride salts: 5.0 mM Mg^{2+}, 1.0 mM Mn^{2+}, and 2.5 mM Co^{2+}. The number of nucleotides per template represents a calculated average, assuming all molecules are copied. In each case, sufficient synthesis was achieved to pass the am3 mutation which is located 83 nucleotides from the Z-5 primer. The reversion frequency was determined from an infectious centers assay of E. coli spheroplasts. The error rate was calculated from the reversion frequency after subtracting the value for uncopied DNA and dividing by the penetrance or frequency of expression (93), which in these experiments was 0.13 [unpublished data of Kunkel and Loeb (94)].

neous AMV DNA polymerase, Mg^{2+}, and equal concentration of nucleotides, the in vitro mutation rate is 1 in 590 ± 300 (95). However, with less purified preparation of AMV DNA polymerase, fewer errors were observed. With E. coli DNA polymerase I, Mg^{2+}, and equal concentrations of nucleotide substrates (30 μM), the reversion rate is 1.4 times that of uncopied DNA and yields a calculated in vitro mutation rate of approximately 1 in 10,000 (94). The calculated error rates for catalysis in the presence of Mn^{2+} and Co^{2+} are higher. Thus Mn^{2+} and Co^{2+} are mutagenic for the synthesis of biologically active DNA in vitro. By determining the sequence of the products of the reaction synthesized in the presence of Mn^{2+} and Co^{2+}, it should be possible to define the specificity of interactions of these metals with the template nucleotides.

15 THE MECHANISM OF GENETIC MISCODING BY METALS

The exact mechanism by which certain divalent metal ions decrease the fidelity of DNA synthesis in vitro is not known. On the basis of the available data, three alternatives can be unambiguously eliminated while three others may still be considered viable mechanisms.

The following three possibilities by which metal ions decrease the fidelity of in vitro DNA synthesis are no longer tenable mechanisms:

1. *Precipitation of Noncomplementary Nucleotides.* It can be argued that the observed increase in error frequency at high metal concentration represents the selective acid precipitation of metal ion complexes containing unincorporated noncomplementary nucleotides. However, physical and enzymatic studies of the products synthesized with AMV DNA polymerase (73), *E. coli* DNA polymerase I (74), and DNA polymerases -α and -β (80) indicate that the noncomplementary nucleotides are incorporated into a polynucleotide chain, predominantly as single-base substitutions. These studies have involved isolation of the newly synthesized product by sedimentation and determination of nearest-neighbor frequencies. Since the latter procedure depends on enzymatic hydrolysis of phosphodiester bonds, it provides unequivocal evidence that the noncomplementary nucleotides are incorporated into a polynucleotide chain.

2. *Metal–Substrate Interactions.* Metal-induced infidelity does not appear to result from selective interactions between particular metals and particular nucleotides. For example, it could be argued that Co^{2+} selectively interacts with the noncomplementary nucleotide and reduces its effective concentration in the reaction mixture. Experiments indicate that this is unlikely. At a high concentration of Co^{2+} (5 mM), the incorporation of dGTP as the complementary nucleotide with a poly(C) template is markedly inhibited. However, at the same Co^{2+} concentration the incorporation of dGTP as a noncomplementary nucleotide with poly[d(A–T)] as the template is undiminished (77). Similar results have been obtained with Mn^{2+} using different DNA polymerases and different template combinations.

3. *Inhibition of "Proofreading" Exonuclease by Metal Ions.* The possibility that decreases in fidelity with divalent metal ions are mediated by inhibition of $3' \rightarrow 5'$ exonucleolytic activity is also unlikely. Eucaryotic DNA polymerases and DNA polymerases from RNA tumor viruses are devoid of such an activity (2), yet mutagenic metal ions decrease the fidelity of these enzymes. Detailed studies on the effect of Mn^{2+} on fidelity, exonucleolytic activity, and monophosphate generation have been carried out with *E. coli* DNA polymerase I. Under conditions in which Mn^{2+} diminishes fidelity, there is no diminution of the $3'$-exonucleolytic activity (78). More importantly, the effect of Mn^{2+} on nucleoside monophosphate generation is opposite to that which one would predict on the basis of a diminished exonucleolytic activity in that the

Possible Mechanisms for Metal-Induced Infidelity

I.   Altered Substrate Conformation

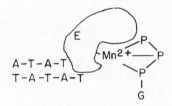

II. Altered Enzyme Conformation

III.   Altered Template Base Specificity

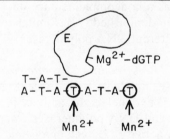

Figure 8   Possible mechanisms for metal ion-induced infidelity of DNA synthesis (see text).

production of noncomplementary nucleoside monophosphates is 40-fold greater with Mn^{2+} than with Mg^{2+} (96).

The decrease in fidelity of metal ions during in vitro DNA synthesis can be explained most directly by any one or more of the types of interactions sketched in Figure 8:

1. *Altered Substrate Conformation.* The ability of Mn^{2+}, Co^{2+}, Ni^{2+}, and possibly Zn^{2+} to substitute for Mg^{2+} as a metal activator focuses (3) on the possibility that the mechanism of change in fidelity by these metals occurs by a substitution at the substrate binding site. Using a variety of DNA polymerases, the frequency of misincorporation at activating concentrations of Mn^{2+} and Co^{2+} is two or three times greater than that observed with Mg^{2+}. Magnetic resonance studies indicate that the interaction of the metal activator involves an enzyme–metal–substrate bridge complex involving the γ-phosphoryl group of the substrate (68). Studies with *E. coli* DNA polymerase I in the absence of template indicate that the bound metal changes the conformation of the substrate to that of the nucleotidyl unit in double-helical DNA. It has been pointed out that this conformation could reduce the frequency of misincorporation (68). On this basis it is likely that differences in conformation of the bound substrate with different metal activators might account for differences in the fidelity of DNA synthesis, particularly at activating concentrations of these cations. However, the current data are not sufficient to eliminate the possibility that differences in fidelity reflect interactions of metal cations with the template and not the enzyme, even when the metal serves as an activator. Thus the parallel incorporation of complementary and noncomplementary nucleotides at activating metal concentrations could simply indicate that polymerization is the rate-limiting event and that the metal-mediated change in fidelity could be at a site other than the substrate site on the enzyme.

2. *Altered Enzyme Conformation.* The decrease in fidelity observed at inhibiting concentrations of metal activators suggests binding of metals at sites in addition to the catalytically active site. Ancillary binding sites for Mn^{2+} were detected on *E. coli* DNA polymerase I by nuclear magnetic resonance studies (65). The demonstration that nonactivating metal cations alter the fidelity of other DNA polymerases is compatible with this concept. Also, evidence has been presented that Be^{2+}, a nonactivating cation, binds to AMV DNA polymerase directly and diminishes the fidelity of DNA synthesis in vitro (87). Multiple nucleotide binding sites on *E. coli* DNA polymerase I have been inferred from magnetic resonance (65) and kinetic studies (97). Thus interactions of metals or metal—nucleotide complexes at distant sites could change the conformation of the polymerase so as to promote misincorporation. So far, attempts to generate an altered DNA polymerase with diminished fidelity by treatment with denaturing agents and heat have not been successful (98). Additional efforts at selective modification of DNA polymerases by tight binding metal complexing agents are required to further define enzyme sites that alter fidelity.

3. *Altered Template Base Specificity.* The direct interaction of metal

ions with phosphates and bases on polynucleotides has been measured by a number of physical techniques (4). Studies on the interaction of Mn^{2+} with activated DNA template by paramagnetic resonance (68) indicate 5 ± 2 very tight sites and 52 weaker sites having an invariant association constant of 68 μM. The largest decreases in fidelity with Mn^{2+} were observed at much higher concentrations (2 to 5 mM). Weak Mn^{2+} binding sites on E. coli DNA polymerase I have been reported (68). However, it is also possible that very weak binding sites on polynucleotides are responsible for diminished fidelity, and these would not be observed in the magnetic resonance experiments. Eichhorn and collaborators initially observed that metal cations can cause enhanced mispairing upon renaturation of polynucleotides (4). Conceivably the metal ions can directly interfere with complementary base pairing or cause a shift in the keto–enol equilibria of the nucleotide. Recent studies by Murray and Flessel (5) indicate that Mn^{2+} and Cd^{2+} promote mispairing during hybridization of the synthetic templates. Moreover, the mispairing with Mn^{2+} can be demonstrated to occur at millimolar concentrations.

16 CONCLUSION

The results in this chapter indicate that mutagenic metal ions alter the fidelity of DNA synthesis. This has been demonstrated with purified DNA polymerases using both synthetic and natural DNA templates. Correlations observed between alterations in fidelity in vitro and mutagenicity or carcinogenicity in vivo are in accord with the hypothesis that infidelity during DNA synthesis may cause mutations. However, metal cations have many effects in vivo. Considerable evidence will be required to document whether or not alterations in the fidelity of DNA synthesis are causally associated with mutations and malignancy. Irrespective of a defined mechanism, the correlation between alterations in fidelity and mutagenicity and/or carcinogenicity indicates the practicality of using fidelity assays as a screen for evaluating possible mutagens and carcinogens. Since these assays are carried out entirely in vitro in homogeneous systems, it is possible to design experiments to understand how metals alter the fidelity of DNA synthesis.

There is a growing realization that environmental agents can be causally associated with mutations and malignancy. One solution to the problem is the removal of these agents from our environment. The studies outlined in this chapter suggest that such an approach may be simplistic. Trace metals have been demonstrated to be required for life and also to be mutagenic and carcinogenic. One will need a means for making judgments as to what are acceptable environmental levels of particular metals. Such judgments will require an understanding of their requirements for metabolism as well as their toxicity, mutagenicity, and carcinogenic effects.

ACKNOWLEDGMENTS

This work was supported by Grant CA–11524 from the National Institutes of Health and by Grant PC776–80439 from the National Science Foundation. R. A. Zakour was supported by Grant 67–0836 from the National Institutes of Health.

We wish to thank Dr. T. A. Kunkel for permitting us to include the data in Table 8 before publication. The data from this laboratory represents the previous contributions of Drs. D. K. Dube, K. Gopinathan, G. Seal, C. W. Shearman, M. A. Sirover, and L. A. Weymouth.

REFERENCES

1. A. Kornberg, *DNA Synthesis*, W. H. Freeman, San Francisco, 1974.

2. L. A. Loeb, in *The Enzymes*, Vol. X, P. D. Boyer, Ed., Academic Press, New York, 1974, pp. 174–209.

3. M. A. Sirover and L. A. Loeb, *Biochem. Biophys. Res. Commun.*, **70,** 812 (1976).

4. G. L. Eichhorn and Y. A. Shin., *J. Am. Chem. Soc.*, **90,** 7323 (1968).

5. M. J. Murray and C. P. Flessel, *Biochim. Biophys. Acta*, **425,** 256 (1976).

6. T. Boveri, in *The Origin of Malignant Tumors*, Williams and Wilkins, Baltimore, 1929.

7. J. Cairns, *Nature*, **255,** 197 (1975).

8. B. N. Ames, W. E. Durston, E. Yamasaki, and F. D. Lee, *Proc. Natl. Acad. Sci. USA*, **70,** 2281 (1973).

9. C. C. Irving, in *Methods in Cancer Research*, Vol. 7, H. Busch, Ed., Academic Press, New York, 1973, pp. 190–244.

10. E. C. Miller and J. A. Miller, in *The Molecular Biology of Cancer*, H. Busch, Ed., Academic Press, New York, 1974, Chap. 10, pp. 377–402.

11. P. Nowell, in *Chromosomes and Cancer*, J. German, Ed., Wiley, New York, 1974, pp. 267–287.

12. H. C. Pitot, *J. Natl. Cancer Inst.*, **53,** 905 (1974).

13. G. H. Robbins, K. H. Kraemer, M. A. Lutzner, B. W. Festoff, and H. G. Coon, *Ann. Int. Med.*, **80,** 221 (1974).

14. M. Swift, L. Sholman, M. Perry, and C. Chase, *Cancer Res.*, **36,** 209 (1976).

15. A. G. Knudson, *Proc. Natl. Acad. Sci. USA*, **68,** 820 (1971).

16. C. P. Flessel, *Adv. Exp. Med. Biol.*, **91,** 117 (1978).

17. F. W. Sunderman, Jr., *Fed. Proc.*, **37,** 40 (1978).

18. H. Nishioka, *Mutation Res.*, **31,** 185 (1975).

19. S. Venditt and L. S. Levy, *Nature*, **250**, 493 (1974).

20. G. Löfroth and B. N. Ames, *Abstr. Eighth Meeting, Environmental Mutagen Society*, Colorado Springs, Colorado, Feb. 13–16, 1977.

21. L. L. Weed, *J. Bacteriol.*, **85**, 1003 (1963).

22. M. Demerec, B. Wallace, E. Witkin, and E. Bertani, *Carnegie Inst. Wash. Yearb.*, **48**, 156 (1949).

23. D. Brusick, F. Gletten, D. Jaqannath, and U. Weekes, *Mutation Res.*, **38**, 386 (1976).

24. B. Rosenberg, *Adv. Exp. Med. Biol.*, **91**, 129 (1978).

25. F. Oehlkers, *Heredity* (Suppl.), **6**, 95 (1953).

26. G. R. Paton and A. C. Allison, *Mutation Res.*, **16**, 332 (1972).

27. B. C. Castro, W. J. Pieczynski, R. L. Nelson, and J. A. DiPaolo, *Proc. Am. Assoc. Cancer Res.*, **17**, 12 (1976).

28. G. Rohr and M. Bauchinger, *Mutation Res.*, **40**, 125 (1976).

29. L. Komczynski, H. Nowak, and L. Rejniak, *Nature*, **198**, 1016 (1963).

30. H. Tsuda and K. Kato, *Gann*, **67**, 469 (1976).

31. A. Fradkin, A. Janoff, B. P. Lane, and M. Kuschner, *Cancer Res.*, **35**, 1058 (1975).

32. G. Fiskesjo, *Hereditas*, **62**, 314 (1969).

33. L. A., Muro and R. A. Goyer, *Arch. Pathol.*, **87**, 660 (1969).

34. A. Orgel and L. E. Orgel, *J. Mol. Biol.*, **14**, 453 (1965).

35. A. L. Reeves and A. J. Vorwald, *Cancer Res.*, **27**, 446 (1967).

36. A. Mazabryad, *Bull. Cancer*, **62**, 49 (1975).

37. G. C. Heath, M. R. Daniel, G. T. Dingle, and M. Webb, *Nature*, **193**, 592 (1962).

38. W. W. Payne, *Arch. Ind. Health*, **21**, 530 (1960).

39. J. C. Heath and M. Webb, *Br. J. Cancer*, **21**, 768 (1967).

40. A. Haddow, F. J. C. Roe, and B. C. V. Mitchley, *Br. Med. J.*, **1**, 1593 (1964).

41. G. J. van Esch and R. Kroes, *Br. J. Cancer*, **23**, 765 (1969).

42. G. D. Stoner, M. B. Shimkin, M. C. Troxell, T. L. Thompson, and L. S. Terry, *Cancer Res.*, **36**, 1744 (1976).

43. W. C. Hueper, *Arch. Pathol.*, **65**, 600 (1958).

44. F. W. Sunderman, Jr., *Am. J. Pathol.*, **46**, 1027 (1965).

45. A. Furst, in H. L. Cannon, and H. C. Hopps, Eds., *Environmental Geochemistry, in Health and Disease*, Memoir 123, Geological Society of America, Boulder, Colorado, 1971, pp. 109–130.

46. F. W. Sunderman, Jr., P. R. Alpass, J. M. Mitchell, R. C. Baselt, and D. M. Albert, personal communication, 1978.

47. F. W. Sunderman, Jr., R. M. Maenza, P. R. Alpass, J. M. Mitchell, I. Damjanov, and P. J. Goldblatt, *Adv. Exp. Med. Biol.*, **91**, 57 (1978).

48. O. Neubauer, *Br. J. Cancer*, **1**, 192 (1947).

49. A. M. Lee and J. E. Fraumeni, Jr., *J. Natl. Cancer Inst.*, **42**, 1045 (1969).

50. M. G. Ott, B. B. Holder, and H. L. Gordon, *Arch. Environ. Health*, **29**, 250 (1974).

51. E. Gross and D. Alwens, *VIII Int. Kongr. Unfall Med. Berufskrank.*, Leipzig, 1938.

52. R. Doll, L. G. Morgan, and F. E. Speizer, *Br. J. Cancer*, **24**, 623 (1970).

53. E. Pedersen, A. C. Hogetveit, and A. Anderson, *Int. J. Cancer*, **12**, 32 (1973).

54. S. Tokudome and M. Kuratsume, *Int. J. Cancer*, **17**, 310 (1976).

55. P. E. Enterline, *J. Occup. Med.*, **16**, 523 (1974).

56. A. K. Roy-Chowdhury, T. F. Mooney, Jr., and A. J. Reeves, *Arch. Environ. Health*, **26**, 253 (1973).

57. L. Foulds, *Cancer Res.*, **14**, 327 (1954).

58. L. A. Loeb, C. F. Springgate, and N. Battula, *Cancer Res.*, **34**, 2311 (1974).

59. J. W. Drake, *Nature*, **221**, 1132 (1969).

60. A. S. Mildvan and L. A. Loeb, *CRC Crit. Rev. Biochem.*, **6**, 219 (1979).

61. A. S. Mildvan, *Ann. Rev. Biochem.*, **43**, 357 (1974).

62. J. J. Hopfield, *Proc. Natl. Acad. Sci. USA*, **71**, 4135 (1974).

63. D. Brutlag and A. Kornberg, *J. Biol. Chem.*, **247**, 241 (1972).

64. L. M. S. Chang and F. J. Bollum, *J. Biol. Chem.*, **248**, 3398 (1973).

65. J. A. Slater, P. Tamir, L. A. Loeb, and A. S. Mildvan, *J. Biol. Chem.*, **247**, 6784 (1972).

66. H. M. Temin and D. Baltimore, *Adv. Virus Res.*, **17**, 129 (1972).

67. L. A. Loeb, K. W. Tartof, and E. C. Travaglini, *Nature*, **242**, 66 (1973).

68. D. L. Sloan, L. A. Loeb, A. S. Mildvan, and R. J. Feldmann, *J. Biol. Chem.*, **250**, 8913 (1975).

69. C. F. Springgate, A. S. Mildvan, R. Abramson, J. L. Engle, and L. A. Loeb, *J. Biol. Chem.*, **248**, 5987 (1973).

70. B. J. Poiesz, G. Seal, and L. A. Loeb, *Proc. Natl. Acad. Sci. USA*, **71**, 4892 (1974).

71. V. D'Aurora, A. M. Stern, and D. S. Sigman, *Biochem. Biophys. Res. Commun.*, **78**, 170 (1977).

72. N. Battula, D. K. Dube, and L. A. Loeb, *J. Biol. Chem.*, **250**, 8404 (1975).

73. N. Battula and L. A. Loeb, *J. Biol. Chem.*, **249**, 4086 (1974).

74. S. S. Agarwal, N. Battula, and L. A. Loeb, *J. Biol. Chem.*, **254**, 101 (1979).

75. L. A. Loeb, L. A. Weymouth, T. A. Kunkel, K. P. Gopinathan, R. A. Beckman, and D. K. Dube, *Cold Spring Harbor Symp. Quant. Biol.*, **XLIII**, 921, (1979).

76. U. Bertazzoni, M. Stefanini, G. P. Noy, E. Giulotto, F. Nuzzo, A. Falaschi, and S. Spadari, *Proc. Natl. Acad. Sci. USA*, **73**, 785 (1976).

77. M. A. Sirover and L. A. Loeb, *J. Biol. Chem.*, **252**, 3605 (1977).

78. M. A. Sirover, D. K. Dube, and L. A. Loeb, *J. Biol. Chem.*, **254**, 107 (1979).

79. D. K. Dube and L. A. Loeb, *Biochem. Biophys. Res. Commun.*, **67**, 1041 (1975).

80. G. Seal, C. W. Shearman, and L. A. Loeb, *J. Biol. Chem.* **254**, 5229 (1979).

81. T. A. Trautner, M. N. Swartz, and A. Kornberg, *Proc. Natl. Acad. Sci. USA*, **48**, 449 (1962).

82. Z. W. Hall and I. R. Lehman, *J. Mol. Biol.*, **36**, 321 (1968).

83. S. Linn, M. Kairis, and R. Hoilliday, *Proc. Natl. Acad. Sci. USA.*, **73**, 2818 (1976).

84. G. Seal and L. A. Loeb, *J. Biol. Chem.*, **251**, 975 (1976).

85. C. F. Springgate and L. A. Loeb, *J. Mol. Biol.*, **97**, 577 (1975).

86. M. Z. Luke, L. Hamilton, and T. C. Hollocher, *Biochem. Biophys. Res. Commun.*, **62**, 497 (1975).

87. M. A. Sirover and L. A. Loeb, *Proc. Natl. Acad. Sci. USA*, **73**, 2331 (1976).

88. M. Radman, G. Villani, S. Boiteux, M. Defais, P. Caillet-Fauquet, and S. Spadari, in *Origins of Human Cancer*, J. D. Watson and H. Hiatt, Eds., Cold Spring Harbor Laboratory, 1977, pp. 903–922.

89. M. A. Sirover and L. A. Loeb, *Science*, **194**, 1434 (1976).

90. M. Miyaki, I. Murata, M. Osabe, and T. Ono, *Biochem. Biophys. Res. Commun.*, **77**, 854 (1977).

91. D. J. Hoffman and S. K. Niyogi, *Science*, **198**, 513 (1977).

92. L. M. S. Chang, G. R. Cassani, and F. J. Bollum, *J. Biol. Chem.*, **247**, 7718 (1972).

93. L. A. Weymouth and L. A. Loeb, *Proc. Natl. Acad. Sci. USA*, **75**, 1924 (1978).

94. T. A. Kunkel and L. A. Loeb, *J. Biol. Chem.*, **254**, 5718 (1979).

95. K. P. Gopinathan, L. A. Weymouth, T. A. Kunkel, and L. A. Loeb, *Nature*, **278**, 857 (1979).

96. L. A. Loeb, D. K. Dube, R. A. Beckman, and K. P. Gopinathan, in preparation.

97. E. C. Travaglini, A. S. Mildvan, and L. A. Loeb, *J. Biol. Chem.*, **250**, 8647 (1975).

98. L. A. Weymouth and L. A. Loeb, *Biochim. Biophys. Acta*, **478**, 305 (1977).

CHAPTER **4**

Metal Ions
and Transfer RNA

MARTHA M. TEETER

Department of Chemistry, Boston University
Boston, Massachusetts
Department of Biology, Massachusetts Institute of Technology
Cambridge, Massachusetts

GARY J. QUIGLEY and ALEXANDER RICH

Department of Biology, Massachusetts Institute of Technology
Cambridge, Massachusetts

CONTENTS

1 INTRODUCTION

The importance of metal ions in stabilizing the three-dimensional structure of polynucleotides has long been recognized (1, 2). Each nucleotide consists of a negatively charged phosphate group, a ribose sugar, and an aromatic base. A polynucleotide is therefore polyanionic. This molecule can only fold compactly with bends and coils if the negative charges are neutralized, either by specific, strong cation coordination or by nonspecific, weak cation binding. When the polynucleotide folds into its native conformation, metal binding sites are formed by phosphate groups. Even though these groups are often remote from one another in the primary sequence, the three-dimensional structure forms them into electronegative pockets or sheets. Knowledge of the three-dimensional structure of polynucleotides permits a detailed analysis of the binding sites and their role in chain folding.

At present, only one species of crystalline transfer ribonucleic acid (tRNA) has yielded structural information sufficient to permit such analysis. Here we limit our review of the voluminous tRNA ion literature to yeast tRNA[Phe], which has been solved and crystallographically refined. We emphasize features of the metal binding sites that appear to be general for all tRNAs and further discuss the role these sites may play in tRNA and polynucleotide function. In order to focus on the structural features of tRNA itself, we briefly consider function and the structural features conserved among different tRNA molecules. We also consider the evidence for specific cation binding sites in yeast tRNA[Phe], both from crystallographic and solution experiments. This is followed by a detailed discussion of the binding site coordination and the relationship these sites may have to tRNA conformational stability.

Three crystallographic investigations of two crystal forms of yeast tRNA[Phe] have identified coordinated cations (3–5). In two of these studies (3, 4) the ion positions have been refined, and in one (3) the coordinated water molecules have been refined semiindependently of the metal. In this refinement (3) the water molecules remained loosely restrained to reasonable bond angles and lengths. Thus the precise coordination geometry of the metal ions can be evaluated objectively in this case.

2 TRANSFER RNA STRUCTURE AND FUNCTION

Transfer RNA is central to protein synthesis in all living systems. It recognizes a triplet codon of the messenger RNA (mRNA) and provides the amino acid specified by the codon to the growing polypeptide chain.

These events take place in the ribosome, which is constructed of over 50 proteins as well as macromolecular RNA. Each cell contains more than 60 different tRNAs corresponding to the codons for the 20 amino acids involved in protein synthesis. These various tRNAs must be distinguished from one another by numerous enzymes such as the tRNA synthetases which add the appropriate amino acid to the tRNA as well as modifying enzymes which change specific bases in some tRNAs but not in others. On the other hand, all chain-elongating tRNAs are recognized by the transfer factor (EF-Tu), by other modifying enzymes, and by the proteins and nucleic acids in the ribosome binding sites. Proteins frequently recognize the state of the tRNA, such as whether or not it has an amino acid attached to its terminus or whether it is bound to cofactors.

Our understanding of the nature of the recognition interaction between macromolecules suggests that these depend on the shapes of the interacting molecular surfaces, on the distribution of hydrophobic and hydrophilic groups and of positively and negatively charged regions on the surface, and finally on the flexibility of these regions. These recognition features result from the three-dimensional folding of the macromolecule and the ions it binds.

All tRNAs whose primary sequences are known can be folded into a cloverleaf secondary structure (6) as illustrated for yeast phenylalanine tRNA (tRNA^{Phe}) in Figure 1. The various regions of this cloverleaf, both single-stranded loops and double-stranded stems, are named for features common to all tRNAs.

The acceptor stem contains seven base pairs. In the dihydrouridine (D) loop, there are two variable regions (α and β) which bracket the two constant guanine residues, contain one to three nucleotides, usually pyrimidines, and frequently include dihydrouracil residues. The D stem has three or four base pairs. The anticodon stem contains five base pairs. Its loop has seven bases. These include the anticodon triplet which is preceded by a conserved uridine (U33) and followed by a hypermodified base (Y37 for tRNA^{Phe}). The TΨC stem contains five base pairs, and the loop has seven bases. The variable features include the number of bases in the D loop and stem and in the variable loop regions. These determine the tRNA class. One molecule of tRNA has been studied in detail crystallographically (class D_4V_5-phenylalanine from yeast); we can therefore explore the ion-related structural features of it and speculate about the structural features of other classes.

The crystallographic structure of yeast tRNA^{Phe} at 2.5 Å resolution has revealed both the overall shape of the molecule and a number of important tertiary structural features of this polynucleotide (7). The molecule is "L shaped." It has two extended double helical arms which each contain two

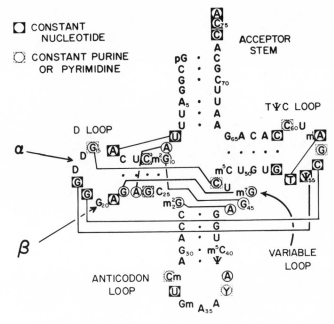

Figure 1 Cloverleaf diagram of yeast phenylalanine tRNA. Bases enclosed in solid or dashed squares are constant for all tRNAs active in chain elongation. Solid lines indicate tertiary interactions between bases and involve one, two, or three hydrogen bonds. The regions α and β in the D loop contain from one to three nucleotides in different tRNA sequences.

of the cloverleaf stems. The horizontal stem at the top of the molecule in Figure 2 contains the acceptor and TΨC stems. The bases ACC in the terminal ACCA are more or less stacked on the stem, with the position of the terminal adenosine less clearly defined. The helix is essentially a normal RNA-11 helix, which absorbs with minimal perturbation the G–U wobble base pair 4–69. There is only a very slight change between base pairs 7–66 and 49–65 where the two stems join. The TΨC stem helix is augmented by the reverse Hoogstein base pairing between T54 and the cationic base m^1A58 from the TΨC loop. Bases Ψ55, C56, and G57 terminate the double-helical stack in a folding arrangement called a "U turn" (8). This feature consists of a sharp turn at the phosphate of 56, which (a) places the phosphate of 57 over Ψ55, (b) puts the phosphate of 58 in a position to hydrogen bond to N-1 of Ψ55, and (c) permits hydrogen bonding from the O-2'-hydroxyl of Ψ55 to the N-7 position of G57. These residues, as well as 54–58 and the terminal pair of the TΨC stem, 53–61, are conserved for nearly all tRNAs, suggesting that this structure exists in all tRNAs. We believe a magnesium ion binds at G53 as discussed below.

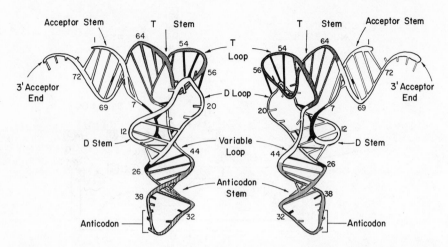

Figure 2 Folding of the ribose phosphate backbone of yeast tRNA^{Phe} is shown as a coiled tube. The numbers refer to nucleotide residues in the sequence. Hydrogen bonding interactions between bases are solid black. Those bases that are not involved in hydrogen bonding to other bases are shown as shortened rods attached to the backbone.

The vertical stem in Figure 2 contains the D and anticodon stems. This stack of bases also terminates in a "U turn," this time in the anticodon loop where the conserved residue U33 plays the same central role as Ψ55 in the TΨC loop. While the anticodon loop does not have a base pair corresponding to 54–58, there is a Mg^{2+} ion found in the center of this loop holding the two sides together, as we discuss in detail below. Thus the "U turn" and the ion hold the anticodon triplet in a position in which they are exposed and can bind to a codon triplet.

Despite the fact that the anticodon stem assumes an RNA-11 double helix, the overall stacking in this region is different from that in the TΨC and CCA acceptor stems. In particular, between the anticodon and the D stems there is a noticeable discontinuity both in the helix axis and in the stacking of the bases. This break takes place at residues m_2^2G26–A44, where there is a purine–purine base pair with a substantial propellarlike twist. The kinking at this point may be related to ion binding at this site as described below.

The most complex region of the molecule, the "core" of the molecule, consists of the four base pairs of the D stem; a portion of the D loop; residues 8 and 9, which connect the D stem to the CCA acceptor stem; and the variable loop (residues 45 to 48), which connects the TΨC and the anticodon stem. The interactions of these four strands form a dense region of base stacking and hydrogen bonding in which there are many base–

base tertiary contacts. These tertiary interactions include hydrogen bonds between G45 and G10; the base triplets A9, U12, and A23 and G13, G22, and m⁷G46; and the base pairs U8–A14 and G15–C48. There are also a number of base backbone interactions, several of which are shown in Figures 3 and 4. The polynucleotide chain in region 8 to 14 forms a sharp bend at the phosphate of 10. This area, together with the backbone of the variable loop, creates a region of high phosphate density. Such a site is favorable for cation binding as noted below. Even though this entire region is tightly packed, it appears possible to construct a similar packing for most tRNAs with a small variable loop. In addition, it is possible to build such a core region, with slight variations, for the large variable-loop tRNAs as well (9).

At the corner of the molecule, bases from the D and TΨC loops interact. This corner is constructed from tertiary base pair interactions between the two conserved guanines (Gs) 18 and 19 of the D loop and residues 55 and 56 of the TΨC loop, which are also conserved and are involved in the TΨC U-turn. The two variable regions of the D loop, α (residues 16 and 17) and β (residue 20), bracket the conserved Gs, bridge back to the remainder of the D loop and create another major bend in the molecule. Here phosphate groups closely approach one another, and two strong magnesium sites are found between residues 17 and 21 as discussed below.

3   CRYSTALLOGRAPHIC INTERPRETATION OF BINDING SITES

A solution of the X-ray diffraction data provides a three-dimensional electron density map. The density peaks in the map are interpreted as specific groups of atoms by using the known connectivity of the polynucleotide chain. However, the positions of ions along the chain are not known. These must be located from a difference electron density map which is calculated by subtracting the diffraction data calculated from the model polynucleotide from the observed diffraction data.

Several criteria must be met before a cation will be observed in the crystallographic experiment. First, it must have a binding constant high enough to occupy a site in a large proportion of the molecules in the crystal (say, 90%). For the conditions used in making the orthorhombic crystals of yeast tRNA^{Phe} (40 mM Mg^{2+}, 15 mg/ml, or 0.56 mM tRNA), the equilibrium binding constant would have to be at least 225. Solution measurements indicate that even weak sites ($K = 10^3$) appear to exceed this association constant (10). Secondly, the site must be ordered, that is, the coordination must be specific. A cation which could occupy two different coordination geometries with the same likelihood would appear as averaged electron density peak and probably be close to the solvent or background electron density. Finally, a cation must have a sufficiently low temperature factor (i.e., vibrational amplitude) in order to be observed. A cation that is coordinated at many points should have a temperature close to the atoms to which it is bound. Regional temperature factors were initially employed and subsequently refined for all the cations (3).

When the conditions cited above have been met, a peak of electron density will be observed. Assignment of this density to specific cations has been generally made on the basis of shape and size. An elongated continuous density was assigned to spermine; spherical, lobed density, to hydrated magnesium ions, although a sodium ion would be quite similar; and small, isolated spheres of density were assigned to water molecules (designated O because hydrogen atoms are not included in the structure). A difference electron density map for spermine and magnesium hydrates can be seen in Figures 3a and 4a. Ideal Mg^{2+}–O distances (2.0 Å) and normal N–C, C–C distances for spermine were assumed, Figures 3b and 4b. The crystal mother liquor contained 3 mM spermine, 20 mM Na^+, 40 mM cacodylate, and 40 mM $MgCl_2$. If sodium ions had been specifically coordinated, we would have expected to see an electron density peak 0.8 Å larger in diameter than the magnesium hydrates. While we cannot rule out sodium hydrates in some of our assignments, the smaller charge-to-size ratio (charge density) of sodium makes it a less likely candidate for assignment to the observed specific coordination sites.

When each cation was assigned to a particular site, the coordination of the cation to the polynucleotide was not initially assumed but was determined by an interpretation of the refined cation positions. Hydrogen bonds were assigned on the basis of reasonable distances and angles.

In the case of the heavy atoms (i.e., ions soaked into the crystal), the original position was normally determined by electron density difference maps calculated using the difference between the scaled derivative and native diffraction data. Since many of these ions were first located in the early stages of the structure solution, the sites were not optimally determined. Accordingly, the derivative data were then rescaled to both the observed and calculated native data and a variety of maps calculated to give a much more accurate view of the binding region. While these sites usually give strong difference densities, the interpretation of them may be limited by the degree of isomorphism of the derivative as well as its resolution. Because of these limitations we have not refined the coordination of these cations.

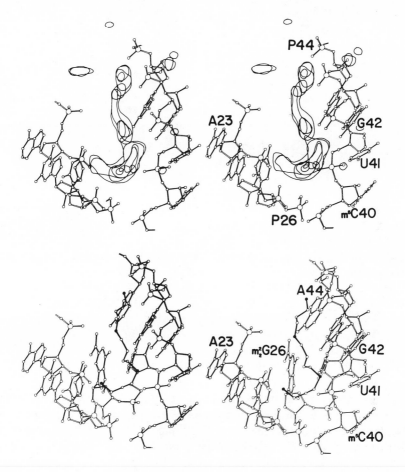

Figure 3 (a) The 2.5 Å difference electron density map is shown together with two segments of backbone in the major groove of the double helix near the juncture of the D stem and the anticodon stem. (b) Interpretation of the electron density map is shown. A spermine molecule with black bonds is shown which goes along the deep groove and then across the groove at the bottom. Two base pairs from the double helix are included in order to show the bottom of the helix. The nitrogen atoms of spermine are black. This figure must be rotated slightly clockwise in order to have the same orientation as the schematic diagram shown on the left side of Figure 2.

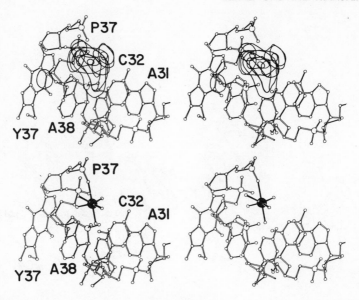

Figure 4 (a) Difference electron density map of the anticodon loop near the last base pair of the anticodon stem. (b) Interpretation of the electron density map showing a magnesium ion (black) with its octahedral coordination. One of the coordinated atoms is an oxygen of phosphate 37. The other coordinated atoms are water molecules that are hydrogen bonded to four neighboring bases.

4   REGIONS OF SPECIFIC CATION BINDING

Maintenance of the double-helical arms of the polyanionic tRNA structure requires the presence of a cationic environment to neutralize the regular array of phosphate groups (1). Weakly binding cations cannot be distinguished from the solvent in the crystallographic experiment. However, specific, presumably strong cation binding can be detected crystallographically.

Specific binding is observed both in sharp bend regions of the "L shaped" molecule and along the double-helical arms. In fact, the ions appear to stabilize many unique features of the tRNA tertiary folding and may be important in regulating conformational changes which occur as tRNA functions in vivo. We describe four regions where binding occurs—the anticodon stem and loop, the corner region, the central core region, and the TΨC and acceptor stem region. Figure 5 provides an overview of most of these binding sites.

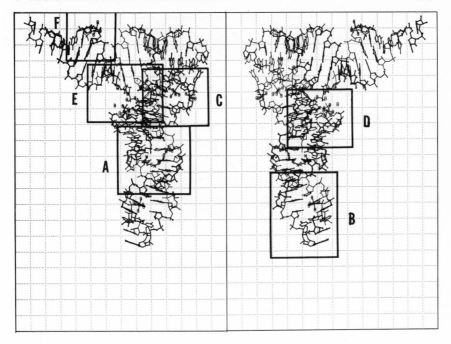

Figure 5 Overview of tRNA atomic model from two sides. Waters and ions are shown with enlarged letters. Boxes show the locations of various regions of ion binding illustrated in subsequent enlarged diagrams. The molecule is shown on a 5 Å grid for reference.

4.1 Anticodon Stem and Loop

Two main binding sites exist in this region (see regions A and B, Fig. 5). In the native orthorhombic crystal form, the first region is occupied by spermine and the second is filled by magnesium. Spermine is among the longest of the naturally occurring polyamines. It is a linear molecule $[NH_3^+(CH_2)_3NH_2^+(CH_2)_4NH_2^+(CH_2)_3NH_3^+]$ with four positively charged nitrogen atoms. When fully extended, it is over 15 Å long. One spermine molecule is found in the deep groove of the double helix. It extends from one end of the D stem into the anticodon stem. The spermine molecule is shaped somewhat like a fish hook. The left end of the NH_3^+ group of the spermine appears to be within hydrogen bonding distance of phosphate 24 and 25 (Fig. 6). The bottom part of the hooked region of the spermine extends across the major groove of the double helix, with a sharp bend occurring at the lower NH_2^+ group which is within 3.1 Å of the O-6 of G42.

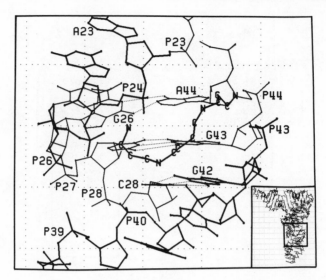

Figure 6 View of the spermine molecule in the major groove between the anticodon and D stems. Inset shows the location of the diagram relative to the whole molecule (A, Fig. 5).

The elongated chain then travels more or less along the groove of the double helix. The upper NH_2^+ group of the spermine is close to phosphate 23, while the NH_3^+ group is within hydrogen bonding distance of phosphates 44 and 43. In this conformation, the spermine appears to be hydrogen bonding to four different phosphate residues on both sides of the deep groove of the double helix.

This spermine molecule has a noticeable effect on the orientation of the anticodon stem. Because of the concentration of positive charges in the major groove, the negatively charged phosphate groups on the opposite sides of the major groove are drawn together. For example, the average distance between phosphates on opposite sides of the major groove at the spermine site is 8.6 Å. This should be compared with an analogous distance across the deep groove in the acceptor TΨC stem of about 12 Å. Thus the spermine residue seems to bring the two strands together, about 3 Å closer than they would be in the absence of the spermine. The helix axes of the acceptor stem and the TΨC stems are approximately colinear, but the helix axis of the anticodon stem deviates approximately 25° from the axis of the D stem. One factor associated with this bend is the propeller twisting of the paired bases m_2^2G26 and A44. However, the presence of a polycation in the deep groove of the double helix undoubtedly makes a major contribution to this bend.

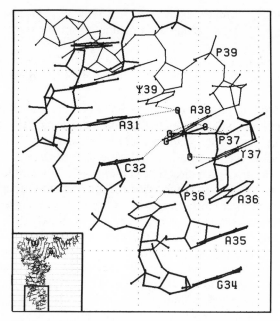

Figure 7   View of the hydrated magnesium ion $Mg(H_2O)_5^{2+}$ in the anticodon loop. The Mg^{2+} is directly coordinated to a phosphate oxygen of residue 37. The waters of hydration show extensive additional hydrogen bonding. Inset shows the location of the diagram relative to the whole molecule (B, Fig. 5).

Binding in this stem region of the monoclinic crystals also occurs with Sm^{3+} (5). This cation bridges the helix between the phosphates of A23 and A44 by means of hydrogen bonds from coordinated water molecules and in addition binds directly to O-6 (2.4 Å) and probably N-7 (3.3 Å) of G45 (the base stacked above A44 of Fig. 6). It is interesting to note that addition of excess Sm^{3+} to crystals in the orthorhombic form (11) broke up the crystals. This could result from a conformational change produced by the Sm^{3+} displacing spermine.

A magnesium ion is located in the upper part of the anticodon loop (Fig. 7). The difference electron density map (Fig. 4b) shows a number of lobes which are due to the coordinated water molecules. The magnesium ion has one oxygen of phosphate 37 in its primary coordination sphere, and the other five oxygen atoms of the coordination sphere are water molecules. The hydrogen bonding scheme of the water molecules in this and other magnesium coordination spheres is similar to that observed by Holbrook et al. (4). The water molecules assume an octahedral coordination and hydrogen bond to residues Y37, A38, Ψ39, A31, and C32 and the phosphate of Y37. This particular coordination of the magnesium ion in the anticodon loop is specific for yeast tRNA^{Phe}. However, it is likely that magnesium ions will also be found in the anticodon loops of the other tRNAs using only slight modifications of the present coordination geometry.

In the monoclinic crystal form of yeast tRNA^{Phe}, Jack *et al.* (5) found that the compound OsO_3py_2 which in solution exists as $Os(VI)O_2(OH)_2py_2$ spans the three bases A29, G30 and A31 by means of hydrogen bonds at the protonated oxygens. Another molecule of this Os(VI) compound as well as cis-$(NH_3)_2PtCl_2$ are directly bound to N-7, O-6, and the phosphate of G34 (5). This site is also on the edge of the anticodon loop rather than within the loop pocket. Its appearance in the monoclinic crystals but not the orthorhombic crystal form is probably due to slightly different packing of the molecules in this region in the two crystal forms.

Sigler and co-workers have found that the reaction of yeast tRNA$_f^{Met}$ with $K_2[OsO_2(OH)_4]$/pyridine (12) produces an Os attachment at a third site in the anticodon region. Formation of this derivative presumably results from the disproportionation of Os(VI) to Os(VIII) and the participation of two pyridine molecules (13). The binding site is again on the outside of the tRNA at a cytosine residue corresponding to position 39 in yeast tRNA^{Phe}. If, in addition to residue 39, the hydroxide from the threonyl at t^{6}A37 is involved, as has been suggested (12), this site would be directly behind the magnesium site illustrated in Figure 7.

4.2 The Corner Region

At the sharp bend in the D loop, two magnesium ions are found coordinated to the three successive phosphate groups of 19, 20, and 21 (see region C, Fig. 5). One magnesium ion has two oxygens from phosphates 20 and 21 in its primary coordination shell (Fig. 8). Furthermore, one coordinated water molecule hydrogen bonds to a phosphate oxygen of residue 17 from a neighboring molecule in the crystal lattice (not shown in Fig. 8 but seen in Fig. 9). This is an important interaction in stabilizing the two-fold screw axis found in both the orthorhombic (14) and monoclinic crystals (15, 16). A second magnesium ion is coordinated to an oxygen from phosphate 19. Water molecules in its first coordination sphere make hydrogen bonds with bases G20, U59, and C60 (Fig. 8) as well as with C56 from another molecule (not shown) which further stabilizes the same screw axis packing. The former magnesium ion can be replaced by a samarium ion (17) (see site B in Fig. 9). It should be noted that in Figures 8 and 9 the tRNA molecule is in two different orientations. These two

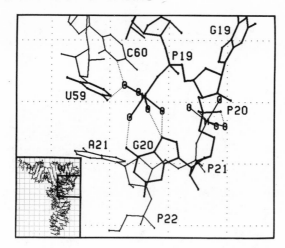

Figure 8 View of a portion of the D loop where two hydrated magnesiums bind. The left magnesium [Mg(H$_2$O)$_5$$^{2+}$] is directly coordinated to a phosphate oxygen of 19, and its waters of hydration appear to be involved in a number of hydrogen bonds. The right magnesium [Mg(H$_2$O)$_4$$^{2+}$] is directly coordinated to one phosphate oxygen each from 20 and 21. Less hydrogen bonding is apparent, although some does take place to a symmetry-related molecule which is not shown. Inset shows the region of the molecule from which diagram is taken (C, Fig. 5).

magnesium binding sites in the D loop are also found in the monoclinic crystal form of yeast tRNA^{Phe} (5). Manganese(II) can replace magnesium ions at the P19 and P20 sites (5).

4.3 The Central Core

The core binding site consists of two regions on either side of the inside bend of the L. In one area (the P10 bend) the close approach of a phosphate chain and a sharp bend produces a high phosphate density (see region D, Fig. 5). This site accommodates both a spermine molecule and a magnesium ion. The other binding site (the P7–P14 region) has a lower phosphate density and consequently contains only loosely bound water molecules (see region E, Fig. 5).

4.3.1 The P10 Bend and Variable Loop.

In this region there is a sharp bend in the polynucleotide chain which occurs at the phosphate of residue 10 at the beginning of the D stem. Two nucleotides, 8 and 9, serve to join the acceptor stem to the D stem (see Fig. 2). Phosphate 10 (P10) is also very close to the extra loop nucleotides (only 7.2 Å from phosphate 47). One spermine molecule is found in a conformation in which it is more or

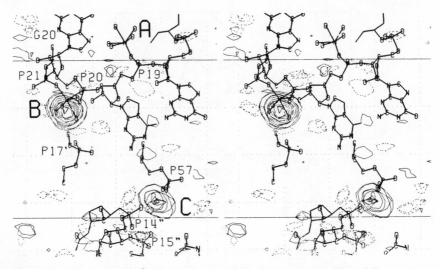

Figure 9 Stereo view of 3 Å difference electron density map showing two of the samarium (Sm^{3+}) sites relative to the determined tRNA structure. Note that one magnesium site, A, shows no indication of being replaced by Sm^{3+}. In a second site, B, the samarium replaces a magnesium but is slightly displaced due to its larger size. In site C there has been no indication of an ordered magnesium, but the presence of a samarium is clear. The samarium at C appears somewhat flattened since it is in a position to coordinate directly to two phosphates on opposite sides. The smaller size of a magnesium would prevent a similar bonding.

less wrapped around P10 such that it is interposed between that chain and the backbone of the extra loop (Fig. 10). The upper NH_2^+ group of the spermine is close to phosphate 47, while the bottom NH_3^+ group is close to ribose 45. In this conformation, the spermine effectively interposes a string of positive charges which neutralize the extended interaction of the two polynucleotide chains involving phosphates 9, 10, and 11 of one chain and phosphates 47, 46, and 45 of the other.

The only metal observed to bind at this site is mercury(II) from hydroxymercurihydroquinone-O,O-diacetate (5). The mercury atom is close to the O-4 of U47 in this variable loop.

The magnesium ion in Figure 10 is hydrogen bonding through its coordination sphere to the four phosphates 8, 9, 11, and 12. It is thus acting in a manner similar to the neighboring spermine residue but in this case neutralizing the charge associated with the very sharp turn of the polynucleotide chain. One of the important derivatives used in solving the structure was obtained by soaking in samarium ions. The magnesium ion of Figure 10 is close to the site of one of the samarium ions (17) and appears to be displaced by it. However, the samarium ion is directly

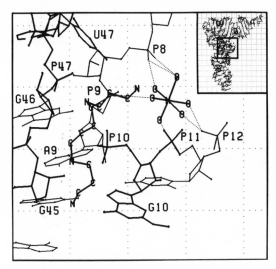

Figure 10   View of a region of the molecule including the P10 bend and part of the variable loop. Here the magnesium hexahydrate [Mg(H$_2$O)$_6$$^{2+}$] bridges the loop formed by the P10 bend, and the spermine molecule coils around P10. Inset shows the region of the molecule from which diagram is taken (D, Fig. 5).

bound to phosphates 8 and 9 and is out of the hydrophilic pocket, as seen in Figure 11. The magnesium ion is held in place by a total of six hydrogen bonds from its coordinated water molecules to neighboring phosphate oxygen atoms (Fig. 10). The compact fit of magnesium into the P10 pocket is illustrated by the stereo van der Waals diagram in Figure 12. It is likely that all tRNA molecules have a tight turn in this region, and a similar ion is likely to be found in other tRNA structures. Indeed, it would be difficult to imagine such a sharp bend of the polynucleotide backbone without neutralization by a cation in that position.

4.3.2   The P7–P14 Region.   Between the two arms of the L-shaped tRNA molecule is a region where the phosphate backbones come fairly close to one another. A conformational change in tRNA may be associated with a flexing of the L-shaped structure. At the inner corner of the L, cations could regulate such a change in structure. As indicated in Figure 13, there are residual peaks of electron density extending from a point between the phosphates of 7, 8, and 14, proceeding up to O-4 of U8 and near the phosphate of 15. It then runs along the phosphate of 16, and down to O-4 and 2'-hydroxyl of U59 and the ribose of C48. This density could conceivably be interpreted as a spermine molecule. However, it is

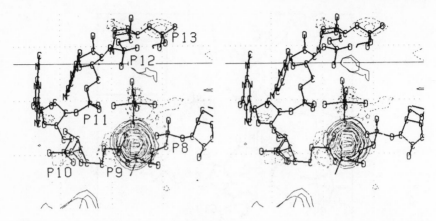

Figure 11 Stereo view of the difference electron density map which shows the samarium ion (Sm^{3+}) that displaces the magnesium ion from the P10 bend. The samarium ion is too large to fit into the P10 pocket and to bridge the two sides of the bend. It does however, because of its large charge-to-size ratio (charge ratio), bind directly to phosphates 8 and 9.

18 Å long and would have to have a coordinated water molecule at one end. Currently, it is interpreted as five water molecules (Fig. 13).

Near one end of this site and between phosphates 7 and 14 is a binding site for Sm^{3+} and Co^{2+} heavy atom derivatives (5). Sm^{3+} (9) binds to the phosphates of 7 and 14 2.5 Å away. Jack et al. (5) have found that Co^{2+} coordinates through water molecules to the phosphates of 7 and 15 and to

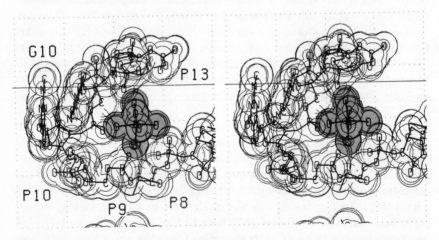

Figure 12 Stereo view of the van der Waals surface of the magnesium ion hexahydrate at the P10 bend of the tRNA. Note the very tight packing of the magnesium ion into this electronegative pocket.

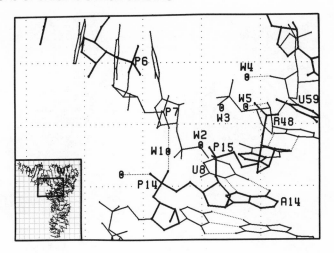

Figure 13 View of a region of the inner corner of the "L" in tRNA. A number of probable water sites appear in this area. Inset shows the location of this region in the overview of the molecule (E, Fig. 5).

N-7 of A14, O-4 of U8, and O-6 of G15. It coordinates directly to N-7 of G15 2.2 Å away.

Ethidium bromide also binds at the 7–14 end of this elongated binding site (18). It stacks over the base pair U8–A14 and near G15 but does not intercalate. The charged pyridium nitrogen in ethidium is 2.1 Å from the phosphate of 14.

A samarium binding site is found in the orthorhombic crystals in the vicinity of P14. (Fig. 9, site C.) It occurs between the phosphate groups of 57 and 14 in two symmetry-related molecules. This intermolecular site is very close to the phosphate of 14. The difference electron density map indicates negative density at P14, which suggests that the phosphate group would have to move to accommodate this cation. Note that the peak at site C is not as intense as the one at site B or the peak in Figure 11. Thus the site is not fully occupied. The fact that phosphate 14 must be moved before samarium can be bound may explain the low occupancy factor in site C.

In another study with the monoclinic crystal form (5), a samarium site was found between the phosphates of 15 and 57 of another molecule. Water molecules have also been observed close to this position (3, 4), between the phosphates of 55 and 57 on the TΨC loop (3, 4) and between N-1 of T54 and the phosphate of 55 (5).

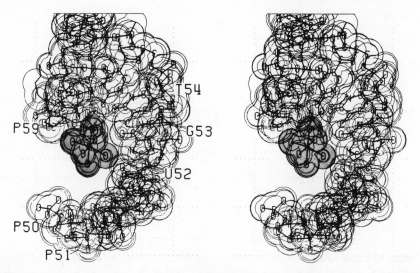

Figure 14 Stereo view of van der Waals surfaces of a magnesium hexahydrate ion binding to the major groove of the TΨC stem. A substantial electronegative pocket is created by a series of phosphates at top and bottom of diagram. The magnesium ion may hydrogen bond through coordinated water molecules to the consecutive bases T54 and G53. The ion does not fit tightly into this pocket.

4.4 TΨC and Acceptor Stem Regions

A number of cations bind in different locations along this helical arm of the tRNA molecule. Although cations in these sites are necessary to allow phosphates to approach one another and form the helix, they are not held as tightly as are the cations found in pockets such as the P10 bend or the anticodon loop. An example of binding in this region is the magnesium site in TΨC helix (Fig. 14). In the deep groove of the TΨC helix, a hydrophilic pocket is created by the bases 52, 53, and 54 and surrounded by the phosphates P59, P60, P50, and P51. Magnesium with its water octahedron fits loosely into the pocket but is only coordinated on one side.

Another magnesium site in this helix occurs at the opposite end, at the 5'-terminal doubly negative phosphate 1 (see region F, Fig. 5). Magnesium is found directly coordinated to the phosphate of 2 and is hydrogen bonding through water molecules to the phosphates of 1 at the 5'-end of the molecule (Fig. 15). Refinement is in progress on this site, and it is possible that the coordination may change.

Two Os derivatives bind in the acceptor stem region, one at the terminal adenosine 76 and the other in the minor groove of the acceptor TΨC

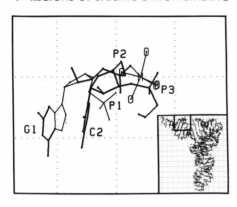

Figure 15 Magnesium binding site between phosphates 1 and 2. Inset shows the region of the molecule from which diagram was taken (F, Fig. 5).

double helix. Interestingly, the site at A76 is a single osmium site (17), while the other site is one of five osmium sites (5). In the former case, OsO_4py_2A in which the Os was attached at the 2′,3′-hydroxyl (19) was reacted with tRNA, and presumably A76 replaced the nucleic acid adenosine. Its coordination is probably similar to that which has been determined crystallographically for OsO_4py_2A (20), Figure 16. Figure 17 shows the prominent features of the Os derivative difference electron density map. A76 is obviously not in the appropriate position to bind to Os at the 2′- and 3′-hydroxyls. However, the negative contours at A76 and at A36 and Y37 from an adjacent molecule suggest that the osmium disturbs the conformation of surrounding bases when it binds. Careful analysis of the native crystal electron densities confirmed that A76 is in the position shown in Figure 17. It is likely that adenosine 76 rotates so that it can

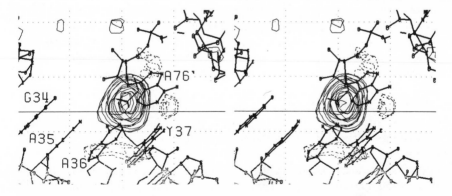

Figure 16 Molecular structure of the bis(pyridine) osmate(VI) ester of adenosine (19). The nitrogen atoms have diagonal shading.

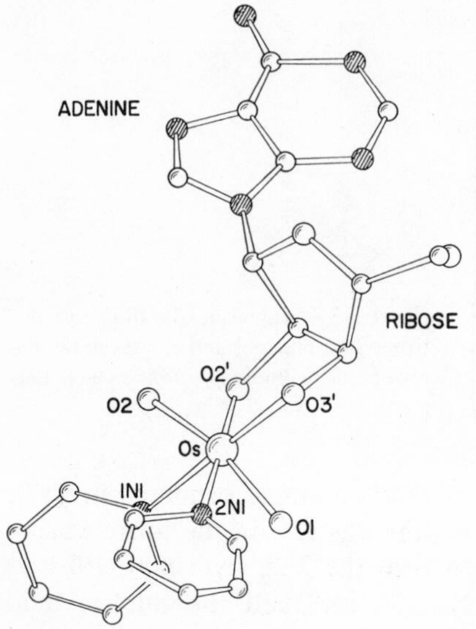

Figure 17 Stereo view of a 4 Å resolution difference electron density map and the tRNA model in the vicinity of the Os(III) binding site. The terminal adenosine is believed to change conformation so as to bind to the Os(III) as shown in Figure 17.

complex with the Os in the manner shown in Figure 16. However, the Os derivative is only isomorphous to 4 Å, suggesting that such a rotation may perturb the native structure considerably when the Os binds. The other Os site occurs at residues 64–66 and has identical coordination to the A–G–A sequence found for the anticodon bases 29–31 (5).

4.5   Solution Evidence for Cation Binding

The importance of cations in the folding of tRNA was demonstrated first in solution by the observation that the ability of denatured tRNA to bind amino acids was restored by the addition of Mg^{2+} (21). A variety of experimental techniques have been used since then to ascertain both the number and location of tRNA metal binding sites. These methods may be divided into those which reveal the general effects of cation binding and those which identified specific sites for strong metal–tRNA binding.

Methods sensitive to changes in tRNA conformation have provided evidence of general tRNA cation binding. In particular, diffusion coefficients derived from light scattering measurements (22), differential migration of native and denatured tRNA on gels (21), and changes in the optical spectra of nucleic acid bases sensitive to general tRNA folding such as the fluorescent Y base and s^4U8 (4-thiouridine) in yeast and *E. coli* tRNAs respectively (10) have been used. All these methods show effects attributable to the presence of added cations. The differences in the order of melting of several regions of the tRNA structure as monitored by the disappearance of the NMR resonance signals (23, 24) as well as changes in the enzyme digestion patterns of tRNA (25) have also been correlated with the presence of cations at varying concentrations.

To detect specific binding in tRNA, a variety of metals have been used as probes to modify the resonance spectrum or the optical spectrum of nucleic acid bases in tRNA. An additional approach has been studies of chemical modification and cleavage of tRNA as a function of metal ion concentration. The nuclear magnetic resonance spectrum of tRNA shows classes of resonances at 11 to 15 ppm and 0 to 3 ppm that have been tentatively assigned to tertiary structure interactions and methyl or methylene protons in the tertiary structure regions, respectively (25, 26). Paramagnetic ions such as Mn^{2+} will broaden these resonances through spin–spin relaxation processes (25, 27). Eu^{3+} produces contact shifts in tRNA resonance signals (21).

Changes in the fluorescence and ultraviolet spectra of the Y base and of 4-thiouridine (s^4U8) in tRNA have provided a particularly valuable probe for metal ion binding in the anticodon and tertiary structure regions. The fluorescence of the Y base is quenched by Eu^{3+} and by Mn^{2+} (25, 29, 30) after three or four metal ions have been added. The fluorescence of s^4U8 is quenched by Eu^{3+}, Tb^{3+}, and even more effectively by Mn^{2+} (28) for both the first and second metal ions added. In addition, Eu^{3+} and Tb^{3+} ions, which are themselves fluorescent, have excitation spectra that are enhanced by tRNA binding. This effect is diminished by photo-induced crosslinking of s^4U8 and C13 (28). The nonfluorescent lanthanides (Sm^{3+}, Gd^{3+}, and Yb^{3+}) as well as Mg^{2+} enhance the s^4U8 fluorescence of s^4U8 (28). Chemical modification and cleavage studies have shown which areas of the tRNA structure are protected by cation interactions and which are accessible to attack by specific reagents (31) or by Pb^{2+} or Zn^{2+} (32, 33).

The presence of metal ions in tRNA has been detected by many methods including atomic absorption for Mg^{2+} (28), radioactive labeling for ^{57}Mn (34), and integration of ESR signals for Mn^{2+} (30). The ionic strength at which these cation binding experiments were conducted has been shown to have had a profound influence on the results obtained. In particular, early binding studies detected cooperative binding (27, 34, 35) at five sites under low salt conditions (below 0.05 M Na^+). This cooperativity was probably due to the folding of the native conformation when the ionic strengths became sufficient to neutralize negative charges (35). Later studies have been conducted at the relatively high salt levels of 0.17 M Na^+ (24, 25, 35, 36) to ensure that tRNA is in its native form. The strong sites appear to be sequential under these conditions.

The number of binding sites presumably due to weak or general binding is 23 ± 5 with binding constants of 10^3 for Mg^{2+} (10). Only four to six strong Mg^{2+} sites with binding constants of 10^5 have been detected (10, 24, 32, 35). This agrees well with the number of Mg^{2+} ions found in the difference Fourier map. Lanthanide ions may have binding constants as high as 10^8 (29). Approximately three to four spermine molecules have been found in the orthorhombic crystals by radioactive labeling experiments (37).

The identification of the location of strong binding sites agrees remarkably well with the crystallographic evidence. In particular, changes in the U8–A14 resonance appear to be associated with the binding of the first Mn^{2+} (25) and the first two Eu^{3+} ions (29). U33 is affected by addition of the fourth cation (29, 30). Changes occur as well in the proton signals assigned to G19–C56, m^1A58–T54, and rT54 itself. Mg^{2+} binding shows changes in the resonances of all of the bases just mentioned in roughly the same order.

One study (38) shows that spermine in the absence of Mg^{2+} stabilizes resonances assigned to m^1A58–T54, U8–A14, and G19–C56. The G19–C56 base pair occurs at the exterior of the molecule and is not accessible to spermine in the crystal because of close intermolecular contacts. There is a Mg^{2+} instead of spermine in the crystal at that site. Subsequent addition of Mg^{2+} affects the U8–A14 and G19–C56 resonances further. Apparently the combination of spermine and Mg^{2+} is more effective than either alone in stabilizing the tRNA structure as seen by increasing the tertiary proton resonances.

5 DISCUSSION

5.1 Nature of Cation Binding Sites

In analyzing the cation sites in tRNA it is useful at first to review the types of metals that are bound and the general evidence for polynucleotide binding by these metals. The metal ions that have been looked at can be divided into three classes: (1) hard metals such as Mg^{2+}, Na^+, and Sm^{3+}; (2) soft metals such as Pt(II), Os(VI), Os(VIII), and Hg(II); and (3) intermediate metals such as Co^{2+} and Mn^{2+}. The hard metals are the alkali, alkaline, and rare earth metals, which have a high charge density. These interact more strongly with the phosphates than with nucleic acid bases. They are known to stabilize the double-helical configuration in DNA (39) and in polyribonucleotides (23), presumably by neutralizing the negatively charged phosphate ions. At low ionic strength, repulsion between phosphate groups unwinds the double helix. Divalent or trivalent cations are much more effective in this neutralization than monovalent ions, as would be expected from their greater charge density, that is, ratio of charge to size. The class 3 intermediate metals are generally first-row transition metals and are more reactive than the hard metals. They are also effective at neutralizing charge and stabilizing the helix at low concentrations; but at high concentrations, they destabilize the helical conformation (1). This destabilization is probably due to the binding of metal ions directly to bases whose electronegative sites are largely buried in the double helix (39). In tRNA, such metal ions coordinate to both phosphate groups and exposed base sites. Finally, the soft, heavy metals (class 2) are the most polarizable of the metals considered and often participate in electron donation with ligands to which they are attached. They bind directly to the bases of polynucleotides. In tRNA, they coordinate most often to N-7 and O-6 of adenine.

A fourth class of cations are polyamines such as spermine, spermidine, and putrescine. The polyamines occur naturally in the cells of many organisms and have been implicated in protein synthesis in general (40) and in tRNA folding in particular (41). They show synergism in combination with Mg^{2+} (38). Spermine facilitates formation of ordered tRNA crystals that yield a high-resolution X-ray diffraction pattern (14). Spermine binds primarily to phosphate groups, but it may also bind to bases.

The cation binding sites observed for tRNA can be divided by shape into two types—electronegative pockets and electronegative clefts. The former surround the cation on three sides with electronegative atoms, for example, Mg^{2+} in the P10 bend (Fig. 12), and the latter hold an elongated cation, namely, spermine in the anticodon stem (Fig. 6). The particular metal ion coordination in each type of site (i.e., to phosphate groups versus to base atoms) depends on the cation itself.

The electronegative pockets we see fall into three groups, corresponding roughly to the three classes of cations that occupy them (Table 1). In group A pockets, phosphate oxygen accounts for all the coordination. These sites bind only class 1 hard metals. Phosphate groups may coordinate to the metal either directly as in the 5'-terminus (one phosphate) or in the P20, P21 site (two phosphates) or indirectly through water molecules as at the P10 bend. All Sm^{3+} ions bind in group A pockets. In group B pockets, both phosphate oxygens and base atoms contribute to the coordination. In general, these sites bind class 3 metals (Mn^{2+}, Co^{2+}) although they appear to be accessible to some class 1 cations as well. The sites at P19 and P20 (Mg^{2+}, Mn^{2+}), at P7 and P14 (Co^{2+}), and in the anticodon loop (Mg^{2+}) are included in this group. Since Mg^{2+} is the major divalent cation present in the cell, the group A and B sites are probably most important in tRNA function. The group C electronegative pockets have

Table 1   Types of Cation Binding Sites in Yeast tRNA[Phe]

Site Classi-fication	Preferred Cations Bound	Coordination	Sites of Each Type[a]
Electronegative pockets			
Group A	Class 1 metals (hard)	Phosphate groups only	P10 bend $Mg(H_2O)_6^{2+}$, $Sm(H_2O)_x^{3+}$ 5′-terminus $Mg(H_2O)_x^{2+}$ P20, P21 $Mg(H_2O)_4^{2+}$, $Sm(H_2O)_x^{3+}$
Group B	Class 3 metals (intermediate)	Phosphate groups and base atoms	P7, P14 $Co(H_2O)_x^{2+}$ P19, P20 $Mn(H_2O)_5^{2+}$, $Mg(H_2O)_5^{2+}$ Anticodon loop $Mg(H_2O)_5^{2+}$
Group C	Class 2 metals (soft)	Base atoms only	U47 Hg(II) in HMHD[b] G34 cis-$(NH_3)_2$ $PtCl_2$, $OsO_2(OH)_2py_2$ A29, G30, A31 $OsO_2(OH)_2py_2$ A64, G65, A66 $OsO_2(OH)_2py_2$
Electronegative clefts	Class 4 cationic polyamines	—	Anticodon/D stem spermine Variable loop/P10 spermine

[a]The symbol $(H_2O)_x$ indicates that the number of water molecules in the hydration shell was not determined.
[b]HMHD is hydroxymercurihydroquinone-O,O-diacetate.

purely base coordination and are occupied by the class 2 soft metals Os[VI] and Pt[II]. Since many of the most reactive heteroatoms in the bases are involved in hydrogen bonding in base pairs (2), the electronegative base atoms that are accessible to the class 2 metals in the native tRNA structure are on the edge of the molecule. Thus these sites might be called more accurately chelation sites rather than electronegative "pockets."

The electronegative clefts can be characterized by closely approaching chains of phosphate groups (P10 bend and variable loop) or phosphates and bases (anticodon stem). These sites are occupied by spermine molecules or by a chain of water molecules (P7 and P14). They may involve fewer well-defined hydrogen bonds than the electronegative pockets. However, the fact that elongated electron density (Fig. 3a) is observed in these sites in the crystal indicates that they are specific enough to satisfy the crystallographic criteria mentioned above.

We would expect all strong cation sites to be observed crystallographically. In addition, a weak site with specific binding should be detected at the Mg^{2+} concentrations present in the crystals, as noted earlier. Such is the case with the Mg^{2+} bound at U52, G53, and T54 (Fig. 14). This ion is loosely confined in an electronegative pocket in a coordination geometry reminiscent of the group C electronegative pocket sites. Other cations in group C pockets are strongly bound because of the reactivity of these soft metals rather than the shape fitting into a highly electronegative pocket (see Fig. 12). As noted above, group C sites are more properly chelation sites. It is difficult to explain why no soft metals bind at the G53 Mg^{2+} site, but perhaps the geometry of the three sequential bases is not favorable.

5.2 Structural Role for Cation Binding

Each region of the tRNA molecule where strong cation binding occurs plays a role in stabilizing the three-dimensional structure and may help to regulate tRNA function. An overview of the strong magnesium and spermine sites is shown in Figure 18.

As noted above, the spermine in the anticodon stem (Figs. 4 and 6) seems to hold the two opposite ribose–phosphate backbones closer together. This is likely to be a significant factor in the 25° deviation from colinearity between the D stem and anticodon stem. In effect, the spermine has produced a bend in the double helix. Because of the long lever arm of the anticodon stem and loop, the actual movement of the anticodon bases relative to the colinear positions induced by the binding of the spermine at this site would be around 10 Å. The anticodon conformation is further stabilized by the magnesium ion which coordinates through water to four bases in the anticodon region. Both the spermine and magnesium binding could be common to all tRNAs, though with modified magnesium coordination in other tRNAs. Both ligands have the effect of substantially immobilizing the anticodon end of the molecule.

It is interesting to consider whether such stabilization has some relevance in terms of the movement of the tRNA molecule within the ribosome during protein synthesis. As aminoacyl tRNA enters the ribosome, it may have spermine and magnesium ions bound to it. It is possible that

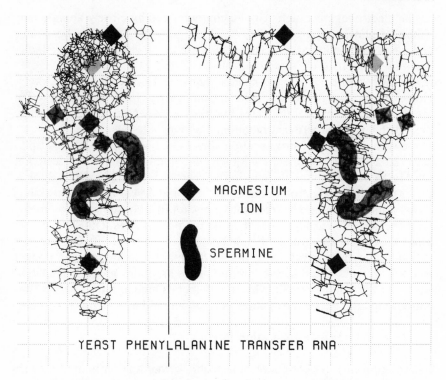

Figure 18 Overview of the ion binding sites to yeast phenylalanine tRNA.

such a stabilization of the anticodon end could facilitate the initial positioning of the aminoacyl tRNA in the A site of the ribosome. However, if translocation of the tRNA from the ribosomal A site to the P site is associated with a conformational change in the anticodon stem and loop, the ribosomal apparatus might accomplish this change by removal of the spermine and magnesium ion from the anticodon end. The stimulatory effect of spermine or other polyamines on protein synthesis in vitro has been well documented. However, further experimentation will be needed to fully assess the role of polyamines in ribosomal activities.

The polyanionic double helix is surrounded by a weak, nonspecific cationic environment (1), as noted above. The result of specific cationic polyamine binding, however, is to introduce a bend in the helix. This bending can be compared to the stabilizing effect of cations on sharp bends in the ribose–phosphate backbone. One can further conceive of the helical bend as an imbalance in close cationic contacts on opposite sides of the double helix.

It has been suggested that spermine binds to DNA in the minor groove and assumes an extended conformation (42, 43). We find that the spermine is in the major groove of the RNA double helix, but, instead of being extended, it is in a slightly coiled conformation so that the net concentration of cations is greater in a particular region.

The fluorescence of the Y base of yeast tRNA is very sensitive to both magnesium and spermine ions; however, magnesium ions are $2\frac{1}{2}$ times more effective (44). These results are readily understandable in view of the close proximity of the magnesium ion to the Y37 base and the fact that the spermine is more remote.

For tRNAs with four or five nucleotides in the variable loop, it is likely that the spermine interaction with phosphate 10 will be a general feature. In a similar way, the magnesium ion coordinating with phosphates 8, 9, 11, and 12 is also likely to be general, as mentioned above. Both of these ions are reasonably close to the residue U8, as shown in Figure 10. In *Escherichia coli* tRNA with s^4U in position 8, varying spermine and magnesium concentrations modify the s^4U fluorescence and its ability to form the UV photo crosslink (45). Further, the two ions together are somewhat more active than either ion alone. Since U8 is near both ions it is reasonable to believe that they can act synergistically in inducing a folding together of the tRNA chain. It is interesting that spermidine is somewhat more effective than spermine, suggesting it may more readily be accommodated in the site near P10. As shown in Figure 10, the lower NH_3^+ does not seem specifically involved in charge neutralization.

The magnesium ions binding to the phosphates of 19, 20 and 21 have both intramolecular and intermolecular coordination. They contribute to the stability of the packing around the two-fold screw axis, which is an important crystallographic feature of both the orthorhombic and monoclinic crystal forms of yeast $tRNA^{Phe}$. The packing interactions around this axis are extensive. The spermine molecules are probably contributing additional stabilization, especially in the anticodon region. These two factors may explain the highly ordered crystals. Although spermine and magnesium ions stabilize other tRNA molecules in solution, it is probably essential to have a detailed fitting together of molecules as seen around the two-fold screw axis in order to have a well-ordered lattice which yields 2 Å resolution in the diffraction pattern.

In the water binding site at P7 and P14, two phosphate groups, one from the acceptor stem (P7) and one from the D stem (P14), closely approach one another. This site is on the inside corner of the L. If a conformational change in the tRNA involved flexing the two arms of the L, this change might be induced by release of the water molecules found at this location.

The magnesium binding site at base 52–54 near the acceptor and TΨC stems might be important in tRNA binding to the ribosome. Oligonucleotide binding studies (46) suggest that in binding to the ribosome rT54, Ψ55 and C56 may dissociate from tertiary interactions with the D loop. Thus some unfolding of the tertiary structure may be involved in binding to the ribosome. Binding to T54 might be important in such a structural change.

The magnesium site at the 5'-phosphate group could be involved in the binding of tRNA both to the ribosome and its synthetase (the enzyme which adds the appropriate amino acid to the 3'-terminal adenosine). Affinity labeling experiments (47) indicate definite locations on the ribosome for CCA and anticodon binding. The sites for attachment of tRNA to the synthetase have been elucidated by crosslinking (48) and chemical modification (49) experiments. It has been shown that the CCA terminus, the D stem, and the anticodon loop are attached to the synthetase in different systems. The specific sites of attachment appear to vary with the amino acid that is added (charged) and with the species. The varied attachment mode could perhaps account for the specificity of the synthetase in charging the tRNA most effectively with the appropriate amino acid. Both ribosomal and synthetase interactions could involve displacements of the 5' Mg^{2+}.

Experiments suggest that tRNA undergoes a magnesium-dependent conformational change when it has aminoacylated (50). A number of studies have been carried out on the effect of spermine on tRNA aminoacyl synthetase activity. Recently, it has been suggested that spermine binding is about 10^4 times stronger than Mg^{2+} binding in the synthetase complex and that under physiologic conditions spermine may be the catalytically more effective cation in promoting aminoacylation (51). It is clear that cations and polyamines are very important in the aminoacylation of tRNA.

5.3   General Role of Cations in Polynucleotide Binding

The effect of cations and cationic polyamines on the tertiary structure of tRNA can readily be extended to other polynucleotide systems. One can imagine both specific and general binding to be of great importance.

Little is known about messenger and ribosomal RNA structure. However, it is possible in certain systems to construct a stem and loop structure in various regions of the RNA (52). Thus it is conceivable that some aspects of tRNA-like tertiary folding will be found as these molecules are studied further. Cations would certainly play an important role in this folding.

It is interesting to note the effect of an extended group of four positive charges on the conformation of the double helix in one arm of the tRNA. The bend created by this tight binding to one side of the helix could facilitate bending in other double-helical nucleic acids such as chromosomal DNA, RNA in its double helical form, and DNA in sperm or viruses where polyamines neutralize almost half the charge (53).

Finally, one would expect weak, nonspecific cationic binding to stabilize the double helix. It has been suggested that removal of metal ions might facilitate formation of single-stranded helical regions in nucleic acids, thus creating local melting sites (1). In DNA such sites could provide the impetus for transcription to begin (copying DNA to make mRNA). This melting could also provide attachment sites for enzymes, nucleases, or repressor molecules.

6 CONCLUSION

In yeast tRNA^{Phe}, electrostatic charge is neutralized either by individual point charges as with magnesium ions or by extended charges as with spermine. The magnesium point charges seem to be most useful in stabilizing portions of the molecule where there is a sharp folding of the polynucleotide chains, as in the loop regions. The more extended polyamines seem to play a slightly different role in neutralizing charges which are found over an extended region and make it possible for two different portions of the polynucleotide chain to come together. As further macromolecular nucleic acid structures are solved, we expect that other examples will be found demonstrating the various ways in which nature uses both discreet and extended cationic structures to maintain the three-dimensional conformation of the nucleic acids.

ACKNOWLEDGMENTS

This research was supported by grants from the National Institute of Health, National Science Foundation, National Aeronautics and Space Administration, and the American Cancer Society. M. M. Teeter was supported in part by an NIH Postdoctoral Fellowship.

REFERENCES

1. G. L. Eichhorn, in *Inorganic Biochemistry,* Vol. 2, G. L. Eichhorn, Ed., Elsevier, Amsterdam, 1973, pp. 1210–1243.

2. L. G. Marzilli, in *Progress in Inorganic Chemistry*, Vol. 23, S. J. Lippard, Ed., Interscience, New York, 1977, pp. 255–378.

3. G. J. Quigley, M. M. Teeter, and A. Rich, *Proc. Natl. Acad. Sci. USA*, **75**, 64 (1978).

4. S. R. Holbrook, J. L. Sussman, R. W. Warrant, G. M. Church, and S. H. Kim, *Nucleic Acids Res.*, **4**, 2811 (1977).

5. A. Jack, J. E. Ladner, D. Rhodes, R. S. Brown, and A. Klug, *J. Mol. Biol.*, **111**, 315 (1977).

6. B. G. Barrell and B. F. C. Clark, *Handbook of Nucleic Acid Sequences*, Joynson-Bruvers, Oxford, England (1974).

7. S. H. Kim, G. J. Quigley, F. L. Suddath, A. McPherson, D. Sneden, J. J. Kim, J. Weinzerl, and A. Rich, *Science*, **179**, 285 (1973).

8. G. J. Quigley and A. Rich, *Science*, **194**, 796 (1976).

9. T. Brennan and M. Sundaralingam, *Nucleic Acids Res.*, **3**, 3235 (1976).

10. R. Römer and R. Hach, *Eur. J. Biochem.*, **55**, 271 (1975).

11. G. J. Quigley, unpublished results.

12. R. W. Schevitz, M. A. Navia, D. A. Bantz, G. Cornick, J. J. Rosa, M. D. H. Rosa, and P. B. Sigler, *Science*, **177**, 429 (1972).

13. J. J. Rosa and P. B. Sigler, *Biochemistry*, **13**, 5102 (1974).

14. S. H. Kim, G. J. Quigley, F. L. Suddath, and A. Rich, *Proc. Natl. Acad. Sci. USA*, **68**, 841 (1971).

15. J. D. Robertus, J. E. Ladner, J. T. Finch, T. Rhodes, R. S. Brown, B. F. C. Clark, and A. Klug, *Nature*, **250**, 546 (1974).

16. C. Stout, J. Mizano, J. Rubin, T. Brennan, S. Rao, and M. Sundaralingam, *Nucleic Acids Res.*, **4**, 2811 (1976).

17. S. H. Kim, G. J. Quigley, F. L. Suddath, A. McPherson, D. Sneden, J. J. Kim, J. Weinzerl, P. Blattman, and A. Rich, *Proc. Natl. Acad. Sci. USA*, **69**, 3746 (1972).

18. M. Liebman, J. Rubin, and M. Sundaralingam, *Proc. Natl. Acad. Sci. USA*, **74**, 4821 (1977).

19. L. R. Subbaraman, J. Subbaraman, and E. J. Behrman, *Bioinorg. Chem.*, **1**, 35 (1971).

20. J. F. Conn, J. J. Kim, F. L. Suddath, P. Blattman, and A. Rich, *J. Amer. Chem. Soc.*, **96**, 7152 (1974).

21. T. Lindahl, A. Adams, and J. R. Fresco, *Proc. Natl. Acad. Sci. USA*, **55**, 941 (1966).

22. T. Olson, M. J. Fournier, K. H. Langley, and N. J. Ford, *J. Mol. Biol.*, **102**, 193–205 (1976).

23. S. Nishimura and G. D. Novelli, *Biochem. Biophys. Res. Commun.*, **11**, 161 (1963).

24. A. Stein and D. M. Crothers, *Biochemistry*, **15**, 160 (1976).

25. Y. H. Chao and D. R. Kearns, *Biochim. Biophys. Acta*, **477**, 20 (1977).

26. B. R. Reid, N. S. Ribeiro, G. Gould, G. Robillard, C. W. Hilbers, and R. G. Shulman, *Proc. Natl. Acad. Sci. USA*, **73**, 2049 (1975).

27. M. Cohn, A. Danchin, and M. Grunberg-Manago, *J. Mol. Biol.*, **39**, 199 (1969).

28. M. S. Kayne and M. Cohn, *Biochemistry*, **13**, 4159 (1974).

29. D. R. Kearns, J. M. Wolfson, and C. R. Jones, 13th Rare Earth Conference, 1977 in *Proceedings: 13th Rare Earth Research Conference, Wheeling, W. Va.*, G. J. McCarthy, Ed., Plenum, New York, 1978.

30. J. L. LeRoy, M. Gueron, G. Thomas, and A. Favre, *Eur. J. Biochem.*, **74**, 567 (1977).

31. D. Rhodes, *Eur. J. Biochem.*, **81**, 91 (1977).

32. C. Weiner, B. Krebs, G. Keith, and G. Dirheimer, *Biochim. Biophys. Acta,* **432**, 161 (1976).

33. P. Wrede, L. Yee, and A. Rich, for *Nucleic Acids Res.,* in preparation.

34. A. H. Schrier and P. R. Schimmel, *J. Mol. Biol.,* **93**, 323 (1975).

35. A. Stein and D. M. Crothers, *Biochemistry,* **15**, 157 (1976).

36. M. Bina-Stein and A. Stein, *Biochemistry,* **15**, 3912 (1976).

37. A. Geller, G. J. Quigley, and A. Rich, *Biochem. Biophys. Res. Commun.,* submitted.

38. P. H. Bolton and D. R. Kearns, *Biochemistry,* **16**, 5729 (1977).

39. R. Thomas, *Trans. Faraday Soc.,* **50**, 304 (1954).

40. L. Stevens, *Biol. Rev. Cambridge Phil. Soc.,* **45**, 1 (1970).

41. J. R. Fresco, B. Adams, R. Ascione, D. Henley, and T. Lindahl, *Cold Spring Harbor Symp. Quant. Biol.,* **31**, 527 (1966).

42. M. Tsuboi, *Bull. Chem. Soc. Japan,* **37**, 1154 (1964).

43. A. M. Liquori, L. Constantino, V. Crescenzi, V. Elia, E. Giglio, R. Puliti, M. de Santis Savino, and V. Vitagliano, *J. Mol. Biol.,* **24**, 113 (1967).

44. B. Robinson and T. P. Zimmerman, *J. Biol. Chem.,* **246**, 110 (1971).

45. F. Pochon and S. S. Cohen, *Biophys. Biochem. Res. Commun.,* **47**, 720 (1972).

46. V. A. Erdmann, M. Sprinzl, and O. Pongs, *Biochem. Biophys. Res. Commun.,* **54**, 942 (1973).

47. H. M. Oen, D. Pellegrini, C. R. Eilat, and C. R. Cantor, *Proc. Natl. Acad. Sci. USA*, **70**, 2799 (1973).

48. H. J. P. Schoemaker, G. P. Budzik, R. Geigé, and P. R. Schimmel, *J. Biol. Chem.,* **250**, 4433 (1975).

49. L. H. Schulman and J. P. Goddard, *J. Biol. Chem.,* **248**, 1341 (1973).

50. R. Potts, M. J. Fournier, and N. C. Ford, *Nature,* **268**, 563 (1977).

51. T. N. E. Lövgren, A. Peterson, and R. B. Loftfield, *J. Biol. Chem.,* **253**, 6702 (1978).

52. G. A. Luoma and A. G. Marshall, *Proc. Natl. Acad. Sci. USA*, **75**, 4901 (1978).

53. C. W. Tabor and H. Tabor, *Ann. Rev. Biochem.,* **45**, 285 (1976).

CHAPTER **5**

Structural Principles of Metal Ion–Nucleotide and Metal Ion–Nucleic Acid Interactions

LUIGI G. MARZILLI and THOMAS J. KISTENMACHER

Department of Chemistry, The Johns Hopkins University, Baltimore, Maryland

GUNTHER L. EICHHORN

Laboratory of Cellular and Molecular Biology, Gerontology Research Center, NIA, NIH, Baltimore City Hospitals, Baltimore, Maryland

CONTENTS

1 INTRODUCTION

The involvement of metal-containing species in genetic information transfer (1), in nucleotide biochemistry (1), and in mutagenesis (2) and carcinogenesis (3) has led to considerable recent efforts at understanding the structural nature of the interaction of metal compounds with nucleotides and nucleic acids (4–11). This activity has gained further impetus with the discovery of metalloantineoplastic agents, such as cis-[Pt^{II} (NH$_3$)$_2$Cl$_2$], which are believed to bind to nucleic acids (12). Another area of widespread interest is the use of the often unique properties of metals as probes of nucleic acid and nucleotide biochemistry (13–16).

Our objective in this review will be to touch on all of the more important structural aspects of the involvement of metal ions and metal complexes in nucleic acid and nucleotide biochemistry. We do not attempt to be comprehensive in our coverage but rather present selected examples. Within the context of this review we focus primarily on structural aspects and, as part of our objective, attempt to illustrate the applications of various approaches in gaining knowledge of structure, both in the solid and in solution. We also try to correlate the accumulated information available on solution and solid-state structure.

A good starting point is the definitive information presently known about metal complexes of small nucleic acid fragments and mononucleotides. The nucleic acids themselves, except for a few tRNAs (17), do not form crystalline material in the usual sense, and much remains to be learned about the detailed structure of metal–nucleic acid interactions. However, some information is now emerging concerning the relationship between the structures of monomeric complexes and complexes of the polynucleotide biopolymers.

Our treatment will begin with a discussion of the relevant chemical and physical properties of nucleic acid constituents. It should be noted that several more extensive reviews on some of the subject matter covered in this chapter have been written. Emphasis in the past has been on solid-state structure (5–7), kinetics (18), thermodynamics and stability constants (19), and solution spectroscopic properties (4). We concentrate on the general principles that have emerged, and the reader is referred to these other reviews for particulars.

1.1 Physical and Chemical Properties of Nucleic Acid Constituents

Nucleic acids are polymers constructed from monomeric nucleotide units. Nucleotides themselves consist of a heterocyclic purine or pyrimidine base which is attached through a glycosyl bond to a pentose sugar which is in turn attached through a phosphoester linkage to a phosphate or a

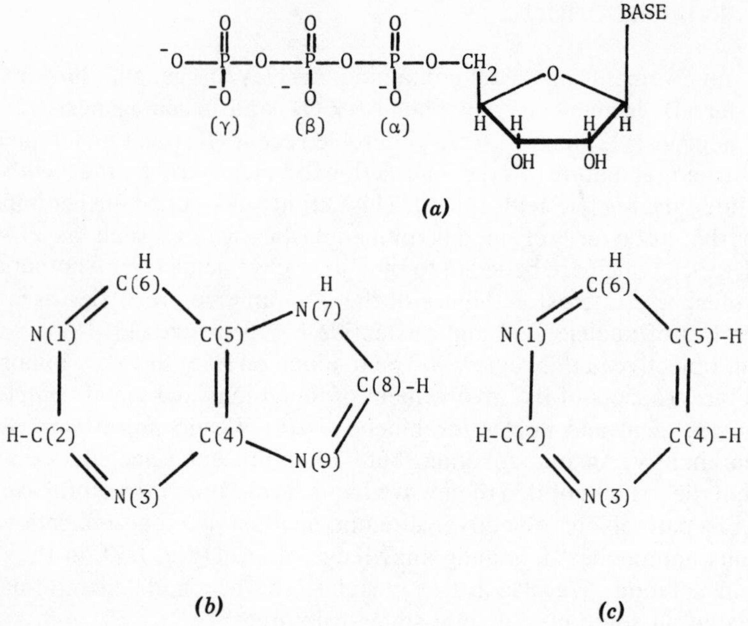

Figure 1 (a) A general nucleoside triphosphate. (b) A schematic representation and the numbering scheme for purine. (c) A schematic representation and the numbering scheme for pyrimidine.

polyphosphate group (Fig. 1). It should be noted that the pyrimidine nucleotides have a six-membered heterocyclic base, while the purine nucleotides have a bicyclic five- and six-membered ring system.

There are also two classes of nucleotides depending on whether the 2′-position on the pentose ring bears a hydroxyl group (ribonucleotides) or not (deoxyribonucleotides). RNAs are polyribonucleotides and DNAs are polydeoxyribonucleotides.

The four most common mononucleotides are illustrated in Figure 2. In the structures presented, the nucleotides are 5′-monophosphates (denoting that the phosphoester linkage is through the 5′-hydroxyl on the pentose ring), and the nucleotide containing the purine adenine is named adenosine 5′-monophosphate. Since the 5′-nucleotides are common monomeric units, they are often abbreviated 5′-AMP, or simply AMP. The corresponding deoxynucleotide of adenine is abbreviated 5′-dAMP, or simply dAMP. The corresponding di- and triphosphates are abbreviated ADP and ATP, respectively. Further common abbreviations employed in nucleic acid chemistry are given in Table 1. In DNA, the commonly occurring bases are T, C, A, and G; whereas in RNA, T is replaced by U.

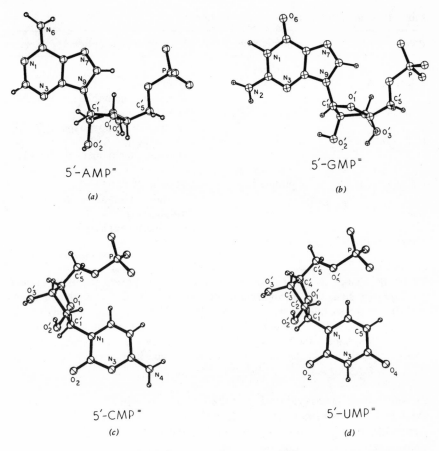

5′-AMP⁼

(a)

5′-GMP⁼

(b)

5′-CMP⁼

(c)

5′-UMP⁼

(d)

Figure 2 Illustrations of the common nucleoside monophosphates: (a) the dianion of 5′-AMP; (b) the dianion of 5′-GMP; (c) the dianion of 5′-CMP: (d) the dianion of 5′-UMP (the complete numbering scheme for the pentose ring is given here). In all cases the conformation about the glycosyl bond is anti.

We have more to say about the properties of polynucleotides later in the chapter. It should be noted, however, that polynucleotides are linked by phosphoester linkages at the 5′ and 3′ sugar positions. Also, the bases are usually stacked in the ordered forms of the polymers.

For their small size, the physical properties of nucleosides and nucleotides are especially rich. Good sources of relevant background information are the reviews by Ts'o (20) and Marzilli (4). The reader should become acquainted with the numbering scheme for the bases and the pentose sugars, Figures 1 and 2. In particular, it should be noted that

Table 1   Common Names and Abbreviations[a]

Base	Nucleoside	Nucleotide
	Purines	
Adenine (A)	Adenosine (A)	Adenylic acid, adenosine monophosphate (AMP)
Guanine (G)	Guanosine (G)	Guanylic acid, guanosine monophosphate (GMP)
Hypoxanthine (H)	Inosine (I)	Inosinic acid, inosine monophosphate (IMP)
	Pyrimidines	
Cytosine (C)	Cytidine (C)	Cytidylic acid, cytidine monophosphate (CMP)
Uracil (U)	Uridine (U)	Uridylic acid, uridine monophosphate (UMP)
Thymine (T)	Thymidine (T)	Thymidylic acid, thymidine monophosphate (TMP)

[a]Note the change in going from the base to the nucleoside for hypoxanthine (6-oxopurine, a guanine analog lacking the 2-amino group). Also, common names refer to ribonucleosides, except for thymidine, which is used commonly for the deoxyribonucleoside. The ribonucleoside is generally named ribothymidine (see ref. 4). In this chapter, the letters A, G, etc., will usually refer to that base in a larger molecule, such as a nucleoside, or to the nucleoside itself.

the sugar-base glycosyl bond is at the ring site N-1 in pyrimidines and at the site N-9 in purines. Also, the primary metal binding sites are the endocyclic nitrogen atoms N-1 and N-7 on the A, N-7 on G, and N-3 on C. These sites are deprotonated at pH 7.

The heterocyclic bases are essentially planar, with the exocyclic amino groups (in A, G, and C) lying in the plane and the lone pair of electrons delocalized into the π-system of the six-membered rings. The possibility exists for tautomerism between the amino and the imino form (in conjunction with protonation at N-3 in G and C, and N-1 in Ade), but it is now widely accepted that the imino tautomers do not exist in significant concentration in solution. Enol forms could exist for C, U, T, G, and H bases, nucleosides, etc., but again these tautomers are not important. The primary sites for protonation are N-1 in A, N-7 in G, and N-3 in C. These sites correspond to most of the principal metal binding sites, except that N-7 in A is also an important metal binding site. The pK_a values of many derivatives can be found in Ts'o's review (20). Some representative values are given in Table 2.

Table 2   Ionization Constants of Nucleosides and Nucleotides (pK$_a$)^a

	Protonation	Deprotonation
Adenosine	3.5 (N-1)	12.5 (S)
3'-AMP	3.6 (N-1)	5.9 (P)^c
5'-AMP	3.7 (N-1)	6.2 (P)^c
Guanosine	2.1 (N-7)	9.2 (N-1), 12.4 (S)
Inosine	1.2 (N-7)^b	8.8 (N-1), 12.3 (S)
Cytidine	4.2 (N-3)	
Uridine		9.5 (N-3), 12.5 (S)
Thymidine		9.8 (N-3), > 13 (S)
5'-TMP		6.5 (P), 10.0 (N-1)

^{a}Adapted from Ts'o (ref. 20, p. 462).

^{b}May be erroneous (20).

^{c}Values for dissociation of first proton from phosphate < 1; secondary proton dissociation ~ 6 for monophosphates of other bases.

There are several conformational aspects of the geometry of the nucleosides and nucleotides. For example, the puckering of the sugar and the rotational conformer about the N-1–C-1' or N-9–C-1' (glycosyl) bond (anti or syn) have been the subjects of an enormous amount of research. At the present time there have been only a few studies in which these conformational properties have been studied as a function of metal binding. However, the bulk of the evidence for metal-free aqueous solutions indicates that nucleosides and nucleotides have an anti conformation about the glycosyl bond. We shall not go into detail here concerning these geometries except to point out that in the anti conformation, the 2'-OH in ribosyl nucleosides or nucleotides is in a position to form a hydrogen bond to N-3 of purines or O-2 of pyrimidines.

Another important property of nucleosides and nucleotides is the possibility that the bases will stack. Although both purine and pyrimidine nucleosides and nucleotides participate, purine–purine stacking is more favorable than purine–pyrimidine stacking, which in turn is more stable than pyrimidine–pyrimidine interactions. However, PMR data have been interpreted to suggest that the six-membered ring in purines participates to a larger extent in the stacking interaction than does the five-membered ring. This evidence takes the form of greater upfield shifts of six-membered ring protons (H-2) than of five-membered ring protons (H-8) as nucleoside or nucleotide concentration is increased. This result is consistent with stacking because the ring current of an adjacent stacked purine will induce upfield shifts. Pyrimidine rings evidently do not have significant ring currents, but upfield shifts of pyrimidine hydrogens can be observed in solutions containing purine derivatives. Base stacking is commonly observed in the solid and can in some cases be an important feature of the crystal packing in metal complexes of purine and pyrimidine derivatives.

Stacking is not favored in organic solvents, and in some cases it is possible to demonstrate that hydrogen-bonding interactions dominate. Consequently, it is possible to observe hydrogen-bonded G–C pairs (nucleosides) in dimethyl sulfoxide (DMSO) (21). In such solutions, the NH PMR resonances are shifted *downfield*. There are three H bonds between G and C [G NH_2 to C O-2; G N-1 H to C N-3; G O-6 to C NH_2]. Pronounced downfield shifts are not observed for DMSO solutions of A and T (or U), presumably because only two H bonds can be formed [A, NH_2 to T, O-4; and T, N-3 H to A, N-1], for nucleosides.

For dinucleoside monophosphates, where intramolecular stacking interactions are expected to be important, diagnostic features are observed in the ORD spectra. Ulbricht (22), for example, has described the application of Cotton effects, both ORD and CD, to the study of both conformational (syn–anti) properties and stacking interactions of nucleic acid derivatives. The conformation of a nucleoside or nucleotide is a major factor in determining the sign of the Cotton effect. Because the anti conformation is more strongly favored in pyrimidine nucleosides than in purine nucleosides, the Cotton effects in pyrimidine compounds are stronger than in purine compounds. Addition of a phosphate group to the nucleoside generally makes very little difference to the ORD spectrum.

2 METAL BINDING MODES AND METAL BINDING SITES

In this section of the chapter we discuss the predominant structural features which have emerged from the recent extensive studies on metal binding to nucleic acid components. Hodgson (5), Bau (6), and Sundaralingam (7) have provided detailed presentations of the many X-ray crystallographic studies, while Marzilli (4) has described the spectroscopic approaches that have identified metal binding sites in solution.

It is important to stress that in solution many different complexes may coexist. Although X-ray crystallography has been useful in identifying some of the species which can be formed, the frequently small differences in free energy between the various isomers may lead to the isolation of compounds favored in a crystalline lattice, particularly if polymeric species can be formed. The combination of multiple bonding, solvation, hydrogen bonding, and stacking capabilities of these ligands makes complexes of nucleic acid derivatives particularly vulnerable to solid-state effects. Such complications also exist and can be important in solution studies.

Because a multitude of binding modes and binding sites are available, we first summarize the binding modes that have been suggested prior to discussing the results obtained from both spectroscopic and X-ray crystallographic studies. The listing of such modes presented in Table 3 presumes that steric factors are not overwhelmingly important. In this respect we note that N-3 of purine nucleosides are not included as possible binding sites. In most cases the presence of the pentose moiety at N-9 of purine nucleosides and nucleotides will sterically discourage metal binding at N-3. The table is most appropriate for complexes of Cu(II), Hg(II), Pt(II), and Ag(I), metal species that typically form complexes with low coordination numbers.

Before proceeding to specific examples of many of the possible binding modes presented in Table 3, we would like to make a few general comments. First, the reader should recall that N-1 and N-6 H_2 of A, N-3 H and O-4 of T, O-6, N-1 H, and N-2 H_2 of G, and O-2, N-3, and N-4 H_2 of C are sites employed in Watson–Crick interbase hydrogen bonding modes. Metal binding at any of the sites must therefore have some effect on the ability of the bases to form hydrogen bonds of the Watson–Crick type. Some sites [N-1 H of G, N-1 of A, and N-3 H of T or N-3 of C] surely preclude the simultaneous presence of metal binding and a Watson–Crick base pairing; other sites are less restrictive and may only weaken the base pairing. Metal bonding at N-7 of purines could, on balance, strengthen base pairing.

Also, many of the sites noted in Table 3 require deprotonation of the base. However, for some metals, bonding can be so favored that an effective pK_a obtains that is considerably lower than the values quoted in Table 2.

Furthermore, there are several "bonding modes" other than those given in Table 3 that have been observed or suggested. For example, many dimeric compounds, consisting of two nucleosides or nucleotides and two metal centers, have been proposed. "Macrochelates" involving several phosphate oxygens and some ring nitrogens simultaneously bound to a metal have also been postulated. Evidence also exists that stacked complexes involving the interaction of the base portion of an uncomplexed nucleotide with the base portion of a complexed nucleotide can be formed. A coordinated water molecule may hydrogen bond to a ring nitrogen or a phosphate oxygen of a nucleotide which is directly bound to the metal via a phosphate oxygen (or a ring nitrogen). Such "indirect"

Table 3   Possible Metal Binding Modes and Sites (Nucleosides and Nucleotides)

A.	**Monodentate Binding Modes**	
	I. Base sites	
	1. A	N-7, N-1, —N-6H$_2$(high pH)
	2. G and I	N-7, N-1(high pH), —C=O-6
	3. C	N-3, —C=O-2, —N-4H$_2$(high pH)
	4. U and T	—C=O-2, —C=O-4, N-3(high pH)
	II. Sugar hydroxyls	Weak and probably not important
	III. Phosphate oxygens	Highly probable, especially with hard metals
B.	**Bidentate Binding Modes**	
	I. Base site plus exocyclic functional group	
	1. A	N-7, —N-6H$_2$ chelation—a weakly possible mode at high pH, definitive evidence is not available
	2. G and I	N-7, —C=O-6 or N-1(deprotonated), —C=O-6 chelation—highly controversial, frequently speculated about in solution studies, only weakly in evidence from solid state work
	3. C	N-3, —C=O-2 chelation weak, but well established; N-3, —N-4H$_2$ chelation—a weakly possible mode at high pH, definitive evidence is not available
	4. U and T	N-3(deprotonated) and —C=O-2 or —C=O-4 chelation—weak, but examples exist for Hg(II)
	II. N(ring), O(sugar)	No true chelate of this type is known
	III. N(ring), O(phosphate)	Likely, but definitive evidence is lacking
	IV. O, O(phosphate)	Well established, especially for hard metals; tridentate chelate also has strong support
	V. O, O(sugar)	An established bonding mode which is most commonly found at high pH

chelates are also possible when the coordinated hydrogen bond donor is an amine. Furthermore, exocyclic groups are likely to participate in such "indirect" chelates, particularly O-6 and N-6 H$_2$ of N-7-bound purines. It is now clear that favorable hydrogen-bonding interactions of this type are ubiquitous and very important in determining complex structure and stability (10).

In the discussion to follow, we make only passing reference to complexes of the sugar-free bases. Eichhorn (1) has pointed out that frequently the base coordinates via the nitrogen which participates in the glycosyl bond (N-9 in purines and N-1 in pyrimidines). There are some exceptions to this trend, and the structural aspects of complexes of the bases have been treated in detail by Hodgson (5). It should be noted that cytosine usually forms complexes via N-3, as first demonstrated crystallographically by Sundaralingam (23) for Cu(cytosine)$_2$Cl$_2$. This was the first structure relevant to nucleoside/nucleotide structure. The structure of [Co^{III}(ethylenediamine)$_2$adenine$_{-H}$]Cl$_2$, in which deprotonated A is coordinated by N-9, revealed significant interligand hydrogen bonding (24) and led to the prediction, which was confirmed in subsequent studies (10), that interligand hydrogen bonding would play an important role in the coordination chemistry of nucleic acid constituents.

2.1   Metal Binding to Adenosine Derivatives

2.1.1   Structural Studies Illustrating Common Binding Modes (Solid State).
Several X-ray diffraction studies have established that N-7 and N-1 are the primary base-binding sites in biochemically interesting systems. We illustrate in Figure 3, for example, the molecular structure of the complex (glycylglycinato)(aquo)(9-methyladenine)copper(II) (25). The complex is square pyramidal with N-7 of the adenine derivative bound to the Cu(II). In addition, there is an interligand hydrogen bond formed from the exocyclic amino group of the purine base to an axially coordinated water molecule. Moreover, the presence of water-bridged, indirect chelates of this type in a variety of other systems serves to indicate the range

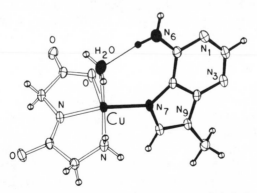

Figure 3   Molecular structure of (glycylglycinato)(aquo)(9-methyladenine)copper(II). (After ref. 25.)

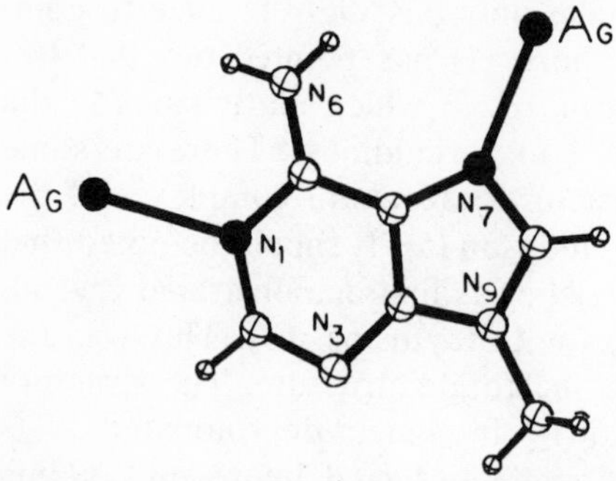

Figure 4 Molecular structure of a fragment of the polymeric structure of the complex formed from $AgNO_3$ and 9-methyladenine. (After ref. 30.)

and importance of such interligand interactions, both with respect to stability of the complexes formed and to specificity of reaction (see below).

Several recent structural studies support also the contention that N-1 of 9-substituted adenine derivatives is an important metal binding site. Skapski (26) has described the structure of [(dichloro)-μ-(9-,methyladenine)cobalt(II)]$_n$ in which both Co(II)–N-7 and Co(II)–N-1 bonds to the substituted adenine molecule are found; Skapski (26) and Taylor (27) have also deduced that the Zn(II)analogue is isomorphous. Furthermore, Taylor (28) was able under acidic conditions to isolate and determine the structure of a Zn(II)–N-1 bonded species, [(trichloro)(9-methyladenine) zincate(II)]$^-$ anion. Lock (29) has also determined the structure of a Pt(II) complex, *trans*-[μ-(9-methyladenine)bis[(dichloro) (diisopropyl-sulfoxide)]platinum(II)], in which Pt(II)–N-7 and Pt(II)–N-1 bonds are formed. Gagnon and Beauchamp (30) have recently studied the reaction between $AgNO_3$ and 9-methyladenine. The complex formed is polymeric with each silver atom diagonally bound to N-1 of one ligand and N-7 of the following ligand; a portion of the structure of this polymeric complex is shown in Figure 4.

In adenine itself, however, N-9 is the most commonly observed site for metal binding, both as a unidentate site and as part of a bridging system in conjunction with N-3. In spite of this, Taylor (31) was able to prepare and study the structure of the complex (trichloro)(adeninium)zinc(II) which shows Zn(II)–N-7 binding. One of the reasons for the preference of N-7 over N-9 in this complex is the presence of an interligand hydrogen bond between one of the chloro ligands and the exocyclic amino group of the adenine ring. Such interligand hydrogen bonding also plays an integral role in most N-7-bound, N-9-substituted adenine complexes.

To return, then, for a moment to the question of interligand hydrogen bonding, we would like to describe briefly several 6-aminopurine

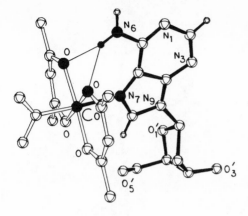

Figure 5 Molecular structure of bis-(acetylacetonato)(nitro)(deoxyadenosine) cobalt(III). (After ref. 32.)

complexes formed in the reaction of purine bases with bis (acetylacetonato)(dinitro)cobalt(III)(32). 6-Aminopurines readily react with this complex resulting in the displacement of one of the nitro groups and the formation of a Co(III)–N-7 bond. The molecular structure of the deoxyadenosine complex is presented in Figure 5. The conformation of the complex is such that the plane of the deoxyadenosine ligand approximately bisects the O–Co–O bonds of two of the acetylacetonato oxygen atoms; in the adopted molecular conformation the exocyclic amino group forms a bifurcated interligand hydrogen bond system. Such an interligand hydrogen-bonding scheme is also present in the triacanthine complex (where triacanthine is 6-amino-3-(γ,γ-dimethylallyl)purine) (33) and the 1,9-dimethyladeninium complexes (34). A comparison of the molecular structures of these three 6-aminopurine complexes, along with that of the 2-aminopurine complex (35), is presented in Figure 6.

The question of metal-phosphate bonding in adenosine nucleotides has received little structural attention at this stage, but the results that have been accumulated are highly significant, particularly with respect to interligand hydrogen bonding. Skapski (36) has determined the structure of the complex [Ni(II)(5'-AMP)(H$_2$O)$_5$], where the metal is bound to N-7 of the 5'-AMP ligand and two of the coordinated water molecules form interligand hydrogen bonds to two of the phosphate oxygen atoms. There is no direct metal-to-phosphate binding. This complex is typical of many metal–nucleotide systems (see following sections, particularly with regard to complexes of 5'-GMP and 5'-IMP) where other ligands about the metal center participate in the formation of interligand hydrogen bonds.

Direct polyphosphate bonding has been established through the very elegant work of Sundaralingam, Cleland, and co-workers. In an attempt to mimic the labile Mg-ATP complexes, these authors have been studying

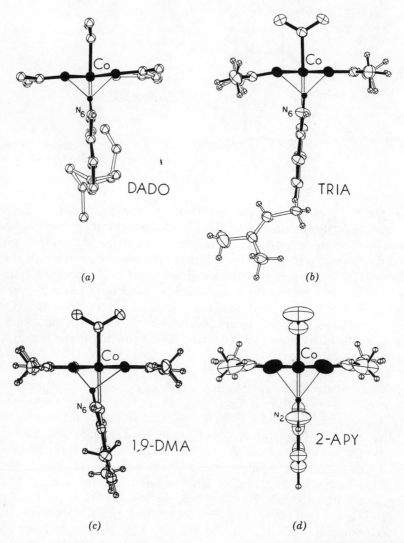

Figure 6 Interligand hydrogen bonding for four bis(acetylacetonato) (nitro) (purine or pyrimidine base)cobalt(III) complexes: (a) the deoxyadenosine complex; (b) the triacanthine complex; (c) the 1,9-dimethyladeninium cation; (d) the 2-aminopyrimidine complex.

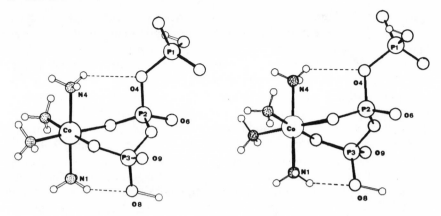

Figure 7 Molecular structure and absolute configuration of (dihydrogentripolyphosphate) (tetraamine)cobalt(III), the Δ enantiomer on the left and the Λ enantiomer on the right. (From ref. 37, with permission.)

the stereochemical properties of inert complexes [Cr(III) and Co(III)] of ATP and their degradation products. Recently, these workers (37) have determined the structure and absolute configuration of (dihydrogen-tripolyphosphate)(tetraammine)cobalt (III), Figure 7, which is a degradation product of one of the biologically active stereomers: β,γ-bidentate (ATP)(tetraammine)cobalt(III). The coordination about the Co(III) center is roughly octahedral, with four ammine ligands and one oxygen atom each from the β and γ phosphates of a tripolyphosphate chain. The adopted conformation of the chelate ring of the tripolyphosphate is stabilized by two intramolecular interligand hydrogen bonds, Figure 7, with two different ammine ligands acting as donors and one terminal phosphate hydroxy and one of the phosphodiester oxygen atoms acting as acceptors. It is believed (37) that such interligand hydrogen bonding is possible in many metal–nucleotide complexes where water (see above, for example) or other hydrogen-bonding ligands complete the metal coordination sphere. Most importantly, it appears that such interligand hydrogen bonding participates in determining which stereomer is biochemically active (37).

The above several studies strongly suggest, then, that interligand hydrogen bonding is a pervading element of metal–6-aminopurine chemistry and biochemistry. The ramifications of such interligand hydrogen bonding extend from complex stability and selectivity to the stabilization of biochemically active metal binding modes.

2.1.2 Solution Studies. Conclusions reached in *careful* solution studies and in the above-mentioned crystallographic studies have been in general agreement, except perhaps that solution studies generally provide more frequent evidence for N-1 involvement in the bonding. Although many diverse techniques can be applied to identify the metal binding site in solution (4), the two most successful approaches have been NMR (4) and Raman spectroscopy (38–43). The deduction of the nature of the binding in solution is not always as definitive as solid-state structure determination. It is essential that great care be exercised in assigning solution structure. This problem is especially difficult for nucleic acid substituents because of the multitude of binding modes suggested in Table 3 and Section 2. In discussion relevant to solution studies, it is our intention to be highly conservative, since many conclusions reached in the literature are frequently lacking justification. In many instances the conclusions are simply wrong. It is now clear that the safest approach to assigning structure is to use a number of spectroscopic techniques which are backed up by preparative and crystallographic studies and by studies incorporating judicious modification of the base, sugar, and/or phosphate moieties. Suggestions (44) that have been made frequently and that in fact have never been convincingly substantiated are (a) that 6-oxopurines form chelates in solution involving N-7 and O-6, and (b) that 6-aminopurines form chelates involving N-7 and the 6-amino group. Evidence for such chelation is sometimes based on changes in infrared stretching frequencies such as the C=O region, but some of us (45) have shown that such changes can be induced by indirect chelation.

Adenine derivatives of all the common nucleic acid bases have received the most attention partly because of the importance of ATP and the availability of two good purine base coordination sites, N-7 and N-1. In addition, the presence of two hydrogen ^{1}H-NMR signals at C-8 and C-2 allows an assessment of both six-membered ring and five-membered ring metal binding (46, 47). The advent of facile ^{13}C-NMR has however made this advantage less important.

One of the early experiments which indicated clearly that both five-membered and six-membered ring sites were involved in the metal binding of adenine derivatives involved ^{1}H-NMR line broadening. Addition of Cu^{2+} salts to adenosine and adenosine nucleotide solutions caused broadening of both the H-8 and H-2 resonances (46). A similar experiment is illustrated in Figure 8. Longitudinal relaxation studies conducted in DMSO (dimethyl sulfoxide), where both the NH and CH resonances can be observed, demonstrated that $1/T_{1p}$ for the 6-NH$_2$ group was large followed by $1/T_{1p}$ for the hydrogens on base carbons and that the sugar H-1' resonance had a small $1/T_{1p}$ (48). These results were in agreement

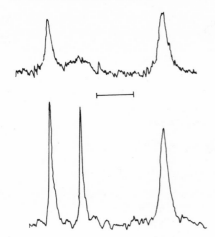

Figure 8 Traces of the PMR spectrum of adenosine (0.2 M, DMSO, 28°, 60 MHz Varian A-60, bar 20 Hz) with no metal present (lower trace) and with (chloro)(N-methyl-N'-salicylideneethylenediamine)copper(II) (1.07 10^{-4} M) present (upper trace). Signals left to right are assigned to the H-8, H-2, and NH_2 resonances. (After ref. 48.)

with line-broadening studies; and, in addition, the six-membered ring site was conclusively identified as N-1 and not N-3. This result is also in excellent agreement with crystallographic studies cited above. In general, the N-1 and N-3 sites on adenine cannot be distinguished as binding sites, but it is usually a reasonable assumption that a substituent at N-9 blocks N-3 coordination. However, in solution studies direct proof of N-1 as opposed to N-3 coordination is seldom obtained.

A number of other studies in solution have shown that a variety of metal species bind to both the six-membered ring (presumably N-1) and the five-membered ring (N-7 sites of N-9-substituted adenine derivatives). A particularly clear case was reported by Theophanides (49) who demonstrated that, at low [Pt^{II}(diethylenetriamine)Cl]/adenosine concentrations, the Pt(II) forms two complexes with Pt(II) on a six-membered ring site in one and on a five-membered ring site in the other. There is little doubt that these sites are N-1 and N-7. At higher Pt(II)/A ratios, a 2Pt:1A complex is formed which most reasonably can be assigned to the N-1, N-7 species similar to a crystallographically established structure (29).

Evaluation of binding sites for diamagnetic metal ions is frequently based on changes in chemical shift. This approach is not always valid, as discussed in the next paragraph, since chemical shifts are difficult to interpret without a large amount of empirical information on hand. Theophanides (49) demonstrated that for Pt(II) a useful aid in assigning the metal binding site was the coupling between ^{195}Pt and the H.

Recent work at Hopkins (33–35) has provided further confirmation of the correlation between solution studies and solid-state structure. The ubiquitous nature of interligand hydrogen bonding involving the exocyclic amino group of 6-aminopurines has been established mainly by crystallo-

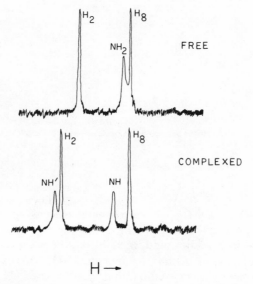

Figure 9 ¹H-NMR traces for a portion of the spectra of free triacanthine and Co(acac)₂(NO₂)(triacanthine).

graphic studies (10,11). Such interactions have been useful in interpreting the stability of complexes in solution, and such reactivity patterns have provided indirect evidence for the existence of intramolecular H-bonding interactions (10, 11). In Figure 9 we present the ¹H-NMR spectra of free triacanthine and triacanthine coordinated to the Co(acac)₂NO₂ moiety (33). A striking feature of the spectrum of the coordinated ligand is that the amino resonance is split into two signals of one proton each. This splitting is understandable in terms of the bifurcated hydrogen bonding scheme illustrated in Figures 5 and 6. The hydrogen bonding in conjunction with coordination of the triacanthine ligand has probably decreased the rate of rotation of the amino group, and the hydrogen bonding has caused a chemical shift nonequivalence of the two proton resonances. This finding provides direct evidence for the interligand hydrogen bonding in solution. A second interesting feature of the spectra in Figure 9 is that the H-2 resonance shifts more than the H-8 resonance, although the metal coordinates to N-7! This result emphasizes the hazard in interpreting chemical shift changes on coordination. The assignment of the H-2 and H-8 resonances in this case was confirmed by specific deuteration studies (33).

Thus far, the emphasis of our discussion of structure in both solution and solid phases has been on the involvement of the binding of the base to the metal. With regard to adenine derivatives, this *emphasis* is based in part on the paucity of solid-state structural information on complexes of larger nucleic acid fragments and of nucleotides. Bau (6) has made a similar observation but was not able to explain the lack of success in obtaining good crystalline derivatives of adenine nucleotides.

In general, nucleosides form much weaker complexes than nucleotides, and the stability of nucleotide complexes is remarkably insensitive to the nature of the metal ion or the nature of the base, except in a few cases involving very soft metals. The reader is referred to earlier reviews (4, 9, 19, 50) and recent work by Sigel (51) for discussion of formation constants. In general, the order of stability is triphosphates > diphosphates > monophosphates >> nucleosides. From the stability trend one would expect many structural reports on nucleotide complexes, but crystallization is usually very difficult.

The nature of the solution phase binding of metal ions to the phosphate groups of nucleotides is almost totally unknown. It is not even clear whether outer- or inner-sphere complexes are formed. Many outer-sphere complexes in which the phosphate group hydrogen bonds to a coordinated water are frequently found crystallographically. The base, the supposedly weaker bonding site, is in fact often coordinated. This finding is exactly opposite to conclusions reached from solution NMR studies (52). Totally outer-sphere binding has been found in the structure of $Ba–AMP \cdot 7H_2O$ (53).

One of the very early and apparently best studies on diamagnetic metal ion binding to phosphate groups was reported by Cohn and Hughes (54). They found that Mg^{2+} ion caused shifts of the β and γ phosphorus-31 resonances of ATP on complexation, indicating that the β and γ phosphate groups were interacting with the metal. Recent work (55) has shown that the shift of the γ ^{31}P resonance can be accounted for by the deprotonation of this phosphate group on coordination since there is a decrease of pK_a of about 1.5. However, a chemically implausible structure was proposed in which the metal ion was bound only through the β phosphate group. A similar study of nucleoside diphosphates led to the conclusion that only the α phosphate group is coordinated to the metal (56).

A recent suggestion (57), with which we agree (58), is that coordination of diamagnetic metal ions, particularly alkaline earth metal ions, may cause such a minor perturbation of even such a sensitive parameter as ^{31}P chemical shift that only minor shifts may accompany coordination. Clearly, some imaginative experiments are needed to finally establish the nature of the metal-to-phosphate binding in solution. Alternatively, better methods of crystallizing nucleotide complexes need to be devised.

One approach, which was mentioned briefly above, is to use inert metal ions that give complexes which can be chromatographically resolved if inner-sphere complexes are formed. Cleland has extensively employed this approach. Since a variety of such inner-sphere complexes are either inhibitors or substrates for a number of enzymes (59), it appears likely that the biochemically significant forms of complexes such as Mg–ATP are inner-sphere. However, a number of possible geometries and conformations can be adopted by these polyphosphate chelates, and different conformers and isomers are specific for any given enzyme. Therefore, there may be a multitude of biologically active Mg–ATP complexes, and one specific geometry of Mg–ATP may not be universally active. Furthermore, the active forms need not be the predominant forms in solution.

A reasonable, although unproved, suggestion as to why the β ^{31}P resonance of nucleoside triphosphates and the α ^{31}P resonance of nucleoside diphosphates is shifted more than the other ^{31}P resonances could involve changes in bond angles (60). Correlations of this type have been made for uncoordinated phosphates (61). However, conceivably outer-sphere conformational changes in the phosphate bond angles could also account for the observed shifts.

The binding of metal ions to ATP has also been investigated using Raman spectroscopy. Changes in the Raman bands attributed to phosphate groups confirm the involvement of these groups in binding to metal ions such as Mg^{2+}, Ca^{2+}, Zn^{2+}, and Mn^{2+} (62–64). The changes with pH in the Raman frequencies of the phosphate band at ~1125 cm^{-1} have two inflection points for Zn^{2+} and Mn^{2+} but only one inflection point for Mg^{2+} and Ca^{2+}. The additional inflection point for Mn^{2+} and Zn^{2+}, which corresponds to deprotonation of the adenine base, led to the suggestion of base binding in these complexes. This conclusion is in general agreement with results obtained using NMR and other spectroscopic approaches.

2.1.3 Unusual Bonding Modes. We have noted earlier in Table 3 that the sugar hydroxyls are possible weak metal binding sites. Good evidence exists for the binding of simple metal ions or complexes to the sugar hydroxyls in solution, but discussion will be deferred to Section 2.4. Conn et al. (65) have determined the structure of the bis(pyridine)osmate(VI) ester of adenosine. In this complex two osmate ester linkages are formed to the ribose group of adenosine and involve the sugar hydroxyls O-2′ and O-3′. Conn et al. (65) believe that the stereochemistry found in this monomeric complex may well serve as a model for one mode of binding of osmium to tRNA; in fact, such an osmate derivative was employed in the phase determination of the structure of tRNA^{Phe} (17).

2.2 Metal Binding to Guanosine and Inosine Derivatives

2.2.1 Structural Studies Illustrating Common Binding Modes (Solid State). There is considerable structural evidence that points to N-7 as the principal metal-to-base binding site in N-9-substituted 6-oxopurines such as guanosine and inosine (5–7). We present, for example, in Figure 10 the molecular structure of the complex (guanosine-5'-monophosphate)penta(aquo)cadmium(II)(66). In this complex, as in several other penta(aquo)metal systems involving 5'-GMP or 5'-IMP, the metal forms a strong bond to N-7 of the base. The presence here of two types of water-bridged indirect chelates involving interligand hydrogen bonding to both the carbonyl oxygen atom O-6 and two of the phosphate oxygen atoms is to be particularly noted. Such indirect chelation, as in the case of the adenosine systems described in Section 2.1, is again an important and general interaction in metal–6-oxopurine chemistry. In fact, Bau (67) found interligand hydrogen bonding of the type —C=O-6 · · · H—N(en)Pt (where en = ethylenediamine) in the complex cation $[Pt^{II}(en)(guanosine)_2]^{2+}$ and has postulated that such interactions may be significant in the antitumor activity of *cis*-$[Pt(ammine)_2X_2]$ complexes. In particular, Bau (67) notes that the antitumor activity of such complexes

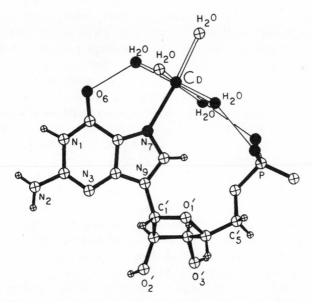

Figure 10 Molecular structure of (guanosine-5'-monophosphate)penta(aquo)cadium(II). (After ref. 66.)

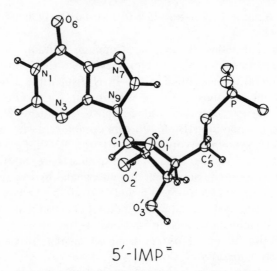

$5'$-IMP$^=$

Figure 11   The dianion of inosine-5′-monophosphate.

parallels the hydrogen bonding ability of the ammine ligand. In contrast to the relatively simple molecular and crystal structure displayed by $Cd^{II}(5'\text{-GMP})(H_2O)_5$, that of the analogous complex with Cu(II), $Cu_3(5'\text{-GMP})_3(H_2O)_8$, is exceedingly complex (68). First, the complex is polymeric with both Cu(II)–N-7 and Cu(II)–O(phosphate) interactions; and secondly, there are three distinct coordination environments about different Cu(II) ions. Two Cu(II) ions have formally identical, but not crystallographically related, square-pyramidal coordination spheres consisting of N-7 of the guanine base, two water ligands, and two phosphate oxygen atoms; while one Cu(II) ion has a square pyramidal environment with N-7, three water molecules, and a phosphate oxygen atom in the primary coordination sphere. We want to mention here that while this complex has an unusually diverse set of coordination environments, the polymeric nature of the material is in accord with many other metal–GMP and metal–IMP complexes (5–7). *Again, in all of these other cases, metal–N-7 and metal–O(phosphate) binding occurs, but in no instance has it been observed that N-7 and the phosphate group of the one nucleotide are coordinated to the same metal center.* While such an N-7 and O(phosphate) intramolecular chelate remains an attractive hypothesis, structural verification has yet to be provided.

One of the most fruitful areas in the study of the binding of metal species to 6-oxopurines has been in the study of metal complexes of inosine 5′-monophosphate (Figure 11). This rare nucleotide is interesting in

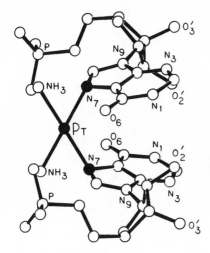

Figure 12 Molecular structure of bis(inosine-5'-monophosphate)(diammine)platinum(II). (After ref. 71.)

itself, as it occurs in the triplet anticodon of a few tRNAs (69). Clark and Orbell (70) were the first to study structurally the metal chemistry of 5'-IMP. The molecular structure of Ni(5'-IMP)(H$_2$O)$_5$ includes a strong Ni(II)–N-7 bond and the triplet of intramolecular hydrogen bonds to the carbonyl oxygen O-6 and two phosphate oxygen atoms as observed in the structurally very similar Cd(II) complex of 5'-GMP, illustrated in Figure 10.

Skapski and Goodgame (71) have described a *cis*-Pt^{II}(NH$_3$)$_2$ complex of 5'-IMP, *cis*-[Pt^{II}(5'-IMP)$_2$(NH$_3$)$_2$]$^{2-}$. In this complex, Pt(II) binds through equatorial sites to two 5'-IMP ligands, which are related about a crystallographic two-fold axis, Figure 12. More recently, Bau et al. (72) have crystallized the complex [Pt^{II}(5'-IMP)$_2$(en)]$^{2-}$, which is isostructural with the compound studied by Goodgame and Skapski. An important and particularly interesting aspect of these complexes is their nonstoichiometry, with the diammine complex having only about 56% of the Pt sites occupied in the crystal and the en complex having only about 36% of the Pt sites occupied. Somewhat surprisingly there are also important differences between these two Pt^{II}–5'-IMP complexes and the [Pt^{II}(guanosine)$_2$(en)]$^{2+}$ complex of Bau (67) and its analog [Pt^{II}-(guanosine)$_2$(NH$_3$)$_2$]$^{2+}$ studied by Cramer (73). We will come back to these interesting compounds in Section 3, on metal binding to polymers.

Before leaving the question of metal binding to 6-oxopurines, we would like to describe some of our work (10) on Cu(II) complexes of the monoanion of theophylline (1,3-dimethyl-2,6-dioxopurine) and to treat very briefly the question of N-7, O-6 chelation in 6-oxopurine systems.

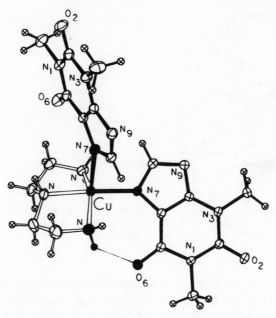

Figure 13   Molecular structure of bis(theophyllinato)(diethylenetriamine)copper(II). (After ref. 74.)

First, we present the structure of the complex Cu^{II}(theophyllinato)$_2$(dien) (74), Figure 13. The coordination environment about the Cu(II) is approximately square pyramidal with, as expected, three of the equatorial sites occupied by the dien chelate and the fourth equatorial site bound to N-7 of the purine base. The interesting aspect of this structure is the presence of a second theophylline monoanion tightly bound to one of the axial sites. The binding of two purines by such different modes remains a unique aspect of this structure. The presence of the interligand hydrogen bond to the equatorially bound purine should also be noted.

In an attempt to address directly the question of N-7, O-6 chelation in 6-oxopurine complexes, we have been studying a series of Schiff base complexes of Cu(II) with the theophylline monoanion (10). In Figure 14 we present the molecular structure of a (N-3,4-benzosalicylidene-N′-methylethylenediamine)Cu(II) complex (75). In this complex the purine is bound through N-7 to an equatorial site. An axially coordinated water molecule and the secondary amino proton on the ethylenediamine terminus of the Schiff base chelate are both involved in interligand hydrogen bonds to the carbonyl oxygen O-6. Substitution of the amino hydrogen on the Schiff base chelate by a methyl group accom-

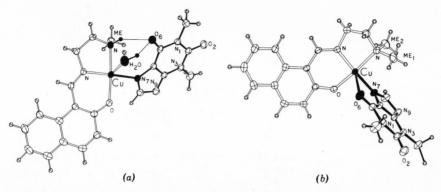

(a)   (b)

Figure 14 Comparison of the molecular structures of the (N-3,4-benzosalicylidene-N'-methylethylenediamine)(theophyllinato)(aquo)copper(II) (A) (after ref. 75) and the (N-3,4-benzosalicylidene-N', N'-dimethylethylenediamine)(theophyllinato)copper(II) (B) (after ref. 76) complexes.

plishes two things: (1) there is no longer any possibility of interligand hydrogen bonding to O-6, and (2) the presence now of the two methyl substituents on the terminal amino group effectively blocks the axial bonding sites to other ligands. In this environment, then, the theophylline monoanion shifts its position relative to the equatorial plane of the complex as found in the monomethyl complex and places O-6 approximately in an axial position (76) (Figure 14). The resulting Cu–O-6 bond at 2.919(3) Å, while hardly overwhelming, is still the shortest such bond observed (5–7). Thus 6-oxopurines will form *weak* N-7, O-6 chelates when no hydrogen bond donor is available, although N-7 obviously remains the primary binding site. Some situations may favor strong N-7, O-6 chelation, such as coordination of N-1-deprotonated guanosine, but to our knowledge definitive X-ray evidence has not been obtained for a strong chelate.

2.2.2. Solution Studies. Both Raman and ^{1}H-NMR studies (4, 43) have recently supported the classic studies of Simpson (77) who showed, with Hg(II), that at low pH metals bound preferentially to N-7 of 6-oxopurine nucleosides, but as the pH was increased, the preferred binding site changed to N-1. At intermediate pH and at sufficiently high metal-to-nucleoside ratios (2 : 1), metal–nucleoside complexes are formed in which the metal binding sites are N-1 and N-7. Some species, Hg^{2+} and CH_3Hg^+, are particularly prone to bind to N-1 in preference to N-7. To this time no X-ray structural evidence has been obtained to support N-1 binding, but there is no doubt as to the formation of such complexes in solution.

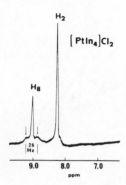

Figure 15 ^{1}H-NMR spectrum of [Pt(Ino)$_4$]Cl$_2$ in D$_2$O. (After ref. 79.)

Numerous studies with paramagnetic metal ions have shown the enhanced relaxation of H-8 as first reported by Eichhorn et al. (4, 78). Coupling between ^{195}Pt and H-8 in the ^{1}H-NMR spectra of Pt(II) complexes of 6-oxopurine nucleosides has been observed (79) (Fig. 15). Additionally, N-1-methylated nucleosides have been shown to form similar complexes to the nonmethylated parents under conditions where N-7 binding has been suggested (42) (Fig. 16).

Tobias's Raman studies (43) have provided some of the best evidence confirming Simpson's conclusions on N-1 binding. From observations of the 1675 cm^{-1} band (D$_2$O) in the Raman spectrum of inosine, the conclusion was drawn that a 1:1 CH$_3$Hg(II)–inosine complex was formed with the Hg attached to N-1 (pH 8, where the predominant free Hg compound is CH$_3$HgOH). The comparable band in 1-methylinosine (1678 cm^{-1}) did not change in intensity when CH$_3$Hg^{II} was added. (Some changes in the

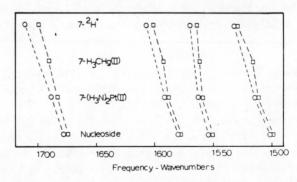

Figure 16   Comparison of the effects of cationic electrophiles, bound at N-7, on the inosine ring modes in the double bond region: (□) inosine; (○) 1-methylinosine. (After ref. 43.)

spectrum of 1-methylinosine were found. These changes result from complex formation at N-7 since exchange of H-8 is enhanced.)

Under the same conditions as the Raman study, the H-2 and H-8 PMR signals of inosine shift upfield, with the H-2 resonance shifting the most for ratios of Hg:inosine up to 1:1. At higher ratios these resonances shift downfield, with the H-8 resonance shifting the most. Furthermore, at these higher ratios H-8 undergoes base-catalyzed exchange with D_2O. The results are eminently consistent with the first Hg binding at N-1 and then a second Hg adding to N-7 (at higher Hg concentrations).

Deprotonation at N-1 leads to a considerable charge delocalization into the ring, and attachment of CH_3Hg^{II} has relatively little effect on this delocalization. Therefore, the binding of one CH_3Hg^{II} enhances the binding of the second CH_3Hg^{II}.

Above, we have addressed the question of N-7, O-6 chelation. It is unlikely that spectroscopic methods will clearly distinguish an N-7, O-6 chelate from an N-7-bound compound. Two groups have isolated either Pt(II) or Pd(II) compounds for which N-7, O-6 chelation has been claimed (80, 81). A chelation mode of binding is one of the ways that the authors can find a satisfactory explanation for the elemental analyses (80, 81). In one of these studies (80), involving a Pt(II) complex, the compound formulated as an N-7, O-6 chelate was formed in the nonaqueous solvent, dimethylformamide, and it is unlikely to be stable in aqueous solutions. We have found, using methods similar to those used to investigate cytidine and cytosine derivatives, that even in nonaqueous solvents the O-6 of guanosine does not appear to interact with hard metal ions such as the lanthanide ions. However, a chelation binding mode might enhance the chances of the interaction of O-6 with metal centers. The considerable activity and speculation on the N-7, O-6 chelation mode of binding derives from suggestions that such chelation may be involved in the reaction of Pt(II) antitumor agents with DNA (82).

2.2.3 Unusual Bonding Modes. In contrast to the lack of "definitive" evidence that metal–N-7, O-6 chelation plays an important role in 6-oxopurine derivatives, there is a substantial literature establishing that 6-mercaptopurines commonly bind metals via an N-7, S-6 bidentate mode. Lippard (83) first demonstrated such a bidentate mode of binding in the complex bis(6-mercapto-9-benzopurine)palladium(II), where the Pd(II)–S distance is about 2.3 Å and can be contrasted to the shortest Cu(II)–O-6 distance of 2.92 Å mentioned in Section 2.2. Sletten (84) has also found a significant N-7, S-6 chelate in the complex (dichloro)(6-mercapto-9-methylpurine)copper(II), where the Cu(II)–S distance is 2.42 Å.

The 6-thio group can also act in a monodentate mode as shown by Caira and Nassimbeni (85) in a dimeric complex of 6-mercaptopurine and Cu^{I}Cl and by Beauchamp (86) in (dichloro)bis(6-mercapto-purine)mercury(II).

There is also some structural evidence that the sugar hydroxyl oxygen atoms O-2' and O-3' can weakly bind a soft metal ion such as Cd(II). Goodgame and Skapski (87) have shown that one of the coordination environments exhibited by a hydrated polymeric complex of Cd(II) and 5'-IMP contains weak chelation by O-2' and O-3'. They believe however that ribose chelation is a result of this particular environment and is probably secondary to metal phosphate binding under other conditions (87).

Finally, Ru(III) can displace the proton at C-8 of caffeine with the formation of a Ru(III)–C-8 bond (88). The deprotonated base can now be considered to be acting as a carbene ligand.

2.3 Metal Binding to Cytidine Derivatives

2.3.1 Structural Studies Illustrating Common Binding Modes (Solid State). From X-ray structural studies it is clear that cytosine shows a high degree of versatility in the binding of metal ions and metal complexes. Early work by Sundaralingam and Carrabine (89) showed that Cu(II) strongly binds to N-3 of cytosine and that there is a significant axial interaction between Cu(II) and the carbonyl oxygen O-2. Such chelation of Cu(II) by N-3 and O-2 of cytosine has been found recently in a variety of structures. We illustrate, for example, in Figure 17 the molecular structure of the complex (glycylglycinato)(cytidine)copper(II) (90) where the Cu(II)–O-2 bond length is 2.74(1) Å. There is also evidence for a weak

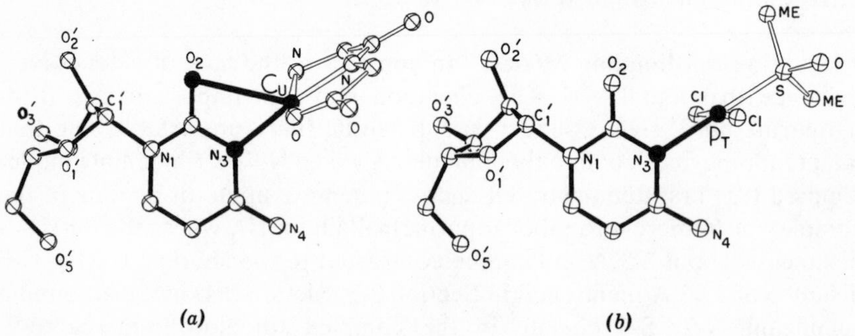

(a)

(b)

Figure 17 Comparison of the molecular structures of (glycylglycinato)(cytidine)copper(II) (a) (after ref. 90) and *trans*-[(dichloro) (dimethyl sulfoxide) (cytidine)platinum(II)] (b). (After ref. 92.)

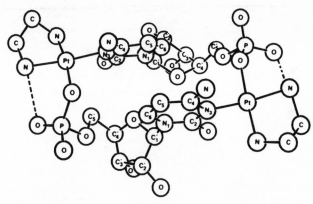

Figure 18 The dimeric unit found in [Pt(II)(en)(5'-CMP)]₂. (From ref. 95, with permission.)

N-3, O-2 chelate in a complex of $HgCl_2$ and 1-methylcytosine (91), where the Hg(II)–O-2 distance is 2.84 Å. Less clear is the role that O-2 plays in the binding of cytosine derivatives to Pt(II) and Pd(II), where the M–O-2 distance is slightly greater than 3 Å. For example, in the complexes *trans*-[(dichloro)(dimethyl sulfoxide)(cytidine)platinum(II)] (92) (Figure 17), *trans*-[(dichloro)(bis(isopropyl) sulfoxide)(1-methylcytosine)platinum(II)] (93), and *trans*-[(dichloro)bis(1-methylcytosine)palladium(II)] (94) strong metal–N(3) bonds are well documented, but the presence of any significant Pt(II)–O-2 or Pd(II)–O-2 interaction is only slightly suggested. Similarly, a strong interaction between Pt(II) and N-3 was found in the 5'-CMP complex Pt(II)(en)(5'-CMP) by Bau and Louie (95). The assemblage of this complex into a dimeric unit via a Pt(II)–O(phosphate) bond underscores the phosphate group of 5'-CMP as a potential metal binding site for Pt(II) (Figure 18).

Furthermore, Aoki (96) has recently determined the structure of an octahedral complex of Mn(II) and 5'-CMP (Figure 19), in which the sole metal binding site to the cytosine base is O-2. Steric constraints imposed by the octahedral environment about the Mn(II) center may preclude the binding of 5'-CMP via N-3, owing to unfavorable interligand interactions (Figure 19). It seems clear however that O-2 can act as a significant unidentate metal binding site as well as acting in concert with N-3 in an N-3, O-2 chelate. An interesting structure of a complex formed between $AgNO_3$ and 1-methylcytosine (97) is discussed in Section 3.

2.3.2 Solution Studies. Many of the complications alluded to above concerning establishing the chelate bonding modes for guanosine in solution also apply to cytosine derivatives. Crystallographic data presented

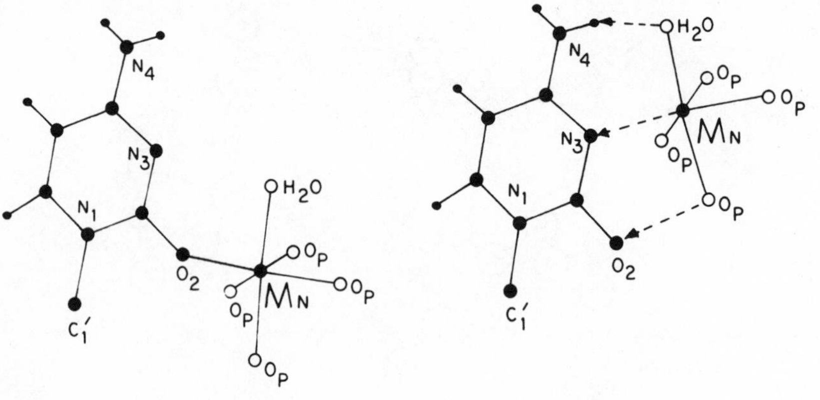

(a)   (b)

Figure 19 (a) Representation of the binding of Mn(II) to O-2 of 5′-CMP. (After ref. 96.) (b) Probable interligand interactions in an octahedral Mn(II) complex of 5′-CMP employing N-3 binding to the cytosine ring.

above have repeatedly indicated some involvement of O-2 in the binding, but clear evidence for such bonding in solution has been difficult to obtain. Binding at N-3, binding at O-2, or binding at both N-3 and O-2 will in general lead to withdrawal of electron density from about the same region of the cytosine ring. Downfield ^{1}H-NMR shifts are expected in all cases. Solution studies of cytosine derivatives have been complicated by the close proximity of the N-3 and O-2 binding sites.

Early line broadening studies could not distinguish these two sites (4, 98), and cautious workers generally concluded that the binding was on either N-3 or O-2 or possibly both. Raman studies comparing the effects of the H^{+}, Pt(II), and Hg(II) compounds clearly indicated that the electrophile was forming its most stable bond with N-3 since the changes induced by these electrophiles had similar patterns (99, 39, 40). All these species are known to bind almost exclusively to N-3 in the solid state.

A second useful approach to assigning binding in these compounds is ^{13}C-NMR. Again, the ^{13}C pattern of shifts is very similar for electrophiles known to bind at N-3 (100, 101). In particular, large upfield shifts are found for C-2 and C-4. There is one puzzling exception to this generalization (102).

Up until recently, a conclusive interpretation of solution data to give spectroscopic criteria for O-2 binding has been lacking. Theoretical studies (103) indicate that the electron density emanating from the O-2 is directed toward almost the same region of space as the electron density emanating from N-3. The unsaturated nature of the ring system makes all

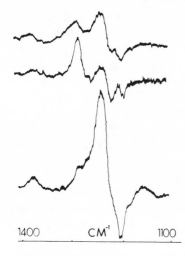

1400   CM^{-1} 1100

Figure 20   Raman difference spectra (in DMSO-d$_6$) obtained by subtracting the Raman spectrum of cytidine 0.7 M in the absence of metal salts from the spectrum of similar solutions containing 0.7 M salt: HgCl$_2$ (bottom), Ba(NO$_3$)$_2$ (middle), and Zn(NO$_3$)$_2$ (top).

of the atoms in the N-3, O-2 region strongly coupled electronically. A delocalized ("allyl type") type of binding to this electron density may be possible but has not been firmly established.

Longitudinal relaxation studies have definitively placed paramagnetic metal ions in the vicinity of N-3 (104, 105). But again, some O-2 involvement cannot be ruled out.

Some metal ions have been reported to cause downfield shifts of the C-2 and upfield shifts of the C-4 resonances in cytidine in DMSO (106). Marzilli et al. have recently reinvestigated this system and have found both Raman and ^{13}C-NMR evidence for O-2 binding in solution (38). In particular, the paramagnetic lanthanides cause large shifts in the C-2 resonance and much smaller shifts of the C-4 resonance. La(III), which is diamagnetic, causes shifts similar to those observed for alkaline earth ions such as Ba^{2+}. In particular, the C-2 resonance is shifted downfield. In addition, the Raman difference spectra for a variety of metal ions complement the ^{13}C spectra. The typically hard metal ions which bind to O-2 cause characteristic changes in the Raman spectra of cytidine (Figure 20). These changes are different from electrophiles known to bind strongly to N-3. The Zn^{2+} ion is a particularly interesting case. Both the Raman and ^{13}C data suggest that this metal binds strongly to both sites. It is not possible to rule out N-3, O-2 chelates where the degree of binding to N-3 and to O-2 is dependent on the metal ion.

It is noteworthy that there are no crystal structures of octahedral metal complexes in which the metal ion is bound to N-3. The hard metal ions have high coordination numbers, and coordination at N-3 has severe steric restrictions. On the other hand, metal ions which form strong complexes with the endocyclic nitrogen of cytosine derivatives are usually soft and have low coordination numbers. A somewhat similar trend extends to the purine bases, but the steric restrictions are less severe.

To further verify the O-2 binding mode, similar studies were performed on dimethylamino-1-methyl cytosine (107). The steric restrictions are even more severe here, and as expected O-2 binding of electrophiles is greatly enhanced over N-3 binding. Similar studies (68) examining the other nucleosides interacting with hard metal ions indicate that the Raman difference spectra and the lanthanide-induced shifts are negligibly small.

It now appears that criteria do exist for distinguishing the N-3 and O-2 binding sites. It is likely that similar criteria for N-7 and O-6 in purines could be found, but the O-6 site appears to be a much weaker binding site than O-2 of cytosine. In fact, it may well be that O-2 of cytosine will prove to be the most important exocyclic binding site in all the common bases.

2.3.3. Unusual Bonding Modes. Lippert and Lock (108) have recently shown under forcing conditions that a metal species can bind at the deprotonated amino group of 1-methylcytosine. It is not anticipated however that a similar reaction will take place under conditions other than those necessary to essentially destroy a polynucleotide.

2.4 Metal Binding to Uridine and Thymidine Derivatives

2.4.1 Structural Studies Illustrating Common Binding Modes (Solid State). There is at present little structural data on metal complexes of uridine and thymidine derivatives. As noted in Section 1.1, these bases contain no nitrogen atom binding sites at pH 7. Soft metal centers such as Hg(II) can weakly bind to O-4 of uracil and dihydrouracil as found in their $HgCl_2$ complexes (109). For most metal centers such as Cu(II), binding to the phosphate group may well be the principal mode of interaction in the absence of deprotonation of the base. Bau and Fischer (110) have demonstrated Cu(II)–O(phosphate) binding in the complex $[Cu^{II}(5\text{-UMP})(2,2'\text{-dypyridylamine})(H_2O)]_2$. Here two Cu(II) ions are bridged by the phosphate oxygen atoms of the 5'-UMP ligand (Fig. 21). Deprotonation at N-3 followed by metal attack at the deprotonated site yields strong metal–N-3 bonding. Stewart (111) studied, for example, the structure of the complex bis-(1-methylthyminato)Hg(II), where two strong linear Hg(II)–N-3 bonds (Fig. 22) as well as a collection of weak intramolecular and intermolecular interactions with O-2 and O-4 of the deprotonated base are found.

Another important metal bonding mode displayed by uridine and thymidine derivatives is the susceptibility of the C-5=C-6 double bond to attack by an MO_4 moiety, where M = Mn, Ru, and Os (4). Base-stabilized

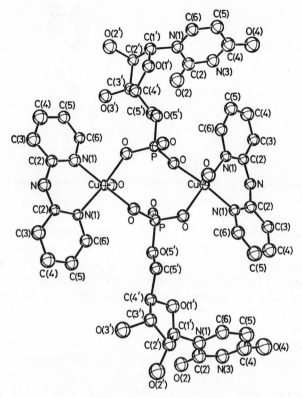

Figure 21 Molecular structure of the ternary complex [Cu(II)(5′-UMP)(2, 2′-dipyridylamine)(H₂O)]₂. (From ref. 110, with permission.)

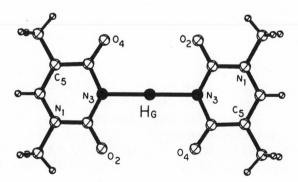

Figure 22 Molecular structure of bis(1-methylthyminato)mercury(II). (After ref. 111.)

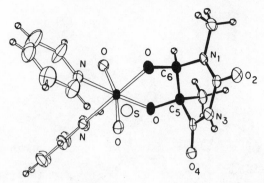

Figure 23 Molecular structure of the osmium tetraoxide bispyridine ester of l-methylthymine. (After ref. 112.)

osmate esters have enjoyed the widest investigation and are important as staining agents in the study of polynucleotide sequence and structure by electron microscopy (4). The structure of the bis(pyridine) osmate ester of 1-methylthymine (112) provides a good model for the mode of binding in the polynucleotides. The molecular conformation of this complex is illustrated in Figure 23. The osmium complex binds as a cis osmate ester to the C-5–C-6 bond of the 1-methylthymine base. The conformation of the 1-methylthymine ester is that of a half-chair with C-6 showing a substantial deviation from the mean plane of the thymine moiety (112).

2.4.2 Solution Studies. Solution studies have confirmed N-3 as the important binding mode under conditions where this site can be deprotonated. Raman (40, 41) and ^{13}C-NMR studies, analogous to those described above for cytidine, have been performed with uridine (4, 113). The effect of a proton on the ^{13}C-NMR spectrum is greater than that of any other electrophile. Thus when electrophiles such as Hg(II) or Pt(II) compounds displace the proton on N-3, there are downfield shifts of the C-2 and C-4 resonances. However, relative to the deprotonated base, these shifts are upfield, as usually found in cytosine derivatives involving N-3 binding.

The problem of distinguishing N-3, O-2, and O-4 binding for metal species or ions other than Hg(II), Pt(II), and Pd(II) complexes is analogous to that described for cytosine derivatives. There are at present no spectroscopic criteria for identifying binding to exocyclic oxygens of uridine or thymidine in solution.

As mentioned above, many metal species are not good competitors relative to the proton for the N-3 site of uridine and thymidine derivatives. Consequently, uridine, which has a cis glycol in the ribose ring, is probably the best nucleoside for studying binding to sugars. The best evidence for metal binding to sugars has been found for DMSO solutions of a variety of Cu(II) salts. A very selective system has been reported by Berger, Tarien, and Eichhorn (114).

Copper(II) acetate broadens the ribose hydroxyl PMR signals in thymidine as well as in ribothymidine (in DMSO), results taken to imply binding of Cu^{2+} to hydroxyl groups in both nucleosides (114). The effects of ribo- and deoxyribonucleosides on the optical spectrum of copper(II) acetate were found to be dramatically different. In DMSO, the acetate-bridged, dimeric structure of copper(II) acetate is preserved. The intensity of the 715 nm peak of copper(II) acetate is greatly diminished by ribonucleosides. Deoxynucleosides produce either no hypochromicity or only a slight decrease in intensity. There is also a spectral enhancement in the near UV induced by ribonucleosides which does not occur with any of the deoxynucleosides studied. Continuous-variation studies established the stoichiometry as 1:1 copper(II) acetate dimer: nucleoside. A structure was proposed in which one of the four acetate bridges is replaced by the 2'- and 3'-hydroxyl groups bridging the coppers, with the preservation of the dimeric structure. The distances involved seemed reasonable from model building. Variation of the sugar using 3'-deoxyadenosine, cytosine arabinoside (in which the 2'- and 3'-OH groups are trans to each other), and triacetyl uridine eliminated the dramatic effect, as expected from the model.

Brun, Goodgame, and Skapski (115) have taken issue with this interpretation. They found that the EPR spectra of *frozen* DMSO solutions of uridine and deoxyuridine do not resemble spectra characteristic of dimers and are essentially identical, except that some unreacted acetate dimer remains in the deoxyuridine solution. This finding prompted them to suggest that this incomplete reaction is responsible for the differentiation between nucleosides and deoxynucleosides. Cytidine appeared to produce two monomeric products, which were thought to be ribose-bound and N-3-bound compounds. Fernando (116) has also argued that the dimeric glycol-bridged structure cannot be correct.

Marzilli and Chalilpoyil (117) have reinvestigated this system and confirmed the spectroscopic results reported from Eichhorn's laboratory. Furthermore, magnetic measurements using the Evans method and *room temperature* EPR spectra clearly show that the product of the reaction between uridine and copper(II) acetate is not a monomer. Acetic acid, not acetate ion, is released during the reaction.

Another type of dimer has been proposed by Kearns (118) for 50/50 $DMSO/H_2O$ solutions containing $Cu(NO_3)_2$ and ribonucleosides to which base has been added. This dimer has been postulated to have two hydroxy bridges with a deprotonated cis glycol completing the coordination sphere of the two coppers. It is different from the species formed from Cu(II) acetate in that there is now one nucleoside per copper(II). Neither species has a fully established structure, but it seems fairly certain that the cis glycol part of the nucleoside is coordinated to the Cu(II). Osmate sugar esters of uridine are also known, and Behrman has reported extensive solution characterization of these compounds (4, 119).

2.4.3 Unusual Bonding Modes. A most interesting class of materials commonly called "platinum-pyrimidine blues" has been the subject of considerable recent interest. It is known that the antineoplastic agent cis-$[Pt^{II}(NH_3)_2Cl_2]$ will react with polyuracil, uracil, thymine, and other pyrimidines to form such materials (120). Much of the recent interest, then, has been induced by the biochemical (antitumor activity and their possible use as electron microscope stains for DNA) as well as the unusual chemical properties of these systems.

Lippard and co-workers (121) have recently been able to crystallize a platinum blue, $[Pt_2(NH_3)_4(\alpha\text{-pyridone})_2]_2(NO_3)_5$, and determine its structure. In the crystal, two cis-$[(NH_3)_2Pt]$ units are bridged by two α-pyridone ligands, with a Pt–Pt separation of 2.78 Å. These dimeric units are assembled across a crystallographic center of symmetry via four interligand hydrogen bonds from the ammine ligands to the exocyclic oxygen atoms of the pyridone ligand and a weaker Pt–Pt interaction at 2.88 Å. Three of the most interesting aspects of this structure are (a) the bridging of two Pt atoms by an N(ring), O(exocyclic) bidentate mode, (b) the overall tetrameric nature of the complex, and (c) the mixed-valence characteristics, with each Pt atom having a formal oxidation state of 2.25.

While more common blues formed from uracil and thymine are probably much higher oligomers, they also appear to be of mixed valence character. Bau and Teo (122) have studied, for example, the EXAFS (extended X-ray absorption fine structure) of a purple and a blue complex formed from the reaction of $Pt^{II}(cyclopropylammine)_2Cl_2$ and uridine. They find that these complexes also contain Pt–Pt interactions between partially oxidized Pt(II) centers.

2.5 Summary and Evaluation of the Importance of Various Binding Modes

All the available endocyclic nitrogen atoms in the purine and pyrimidine nucleosides have been identified as reasonable binding sites, except N-3 of purines. This latter site is blocked by the sugar moiety. Frequently, in solution studies N-3 is often excluded as a potential site without good evidence. Although the N-1 site in purine nucleosides is frequently identified as a binding site in solution, relatively fewer X-ray structural studies have found binding at N-1 for adenosine, and no X-ray structures have been reported for N-1 of inosine or guanosine. The exocyclic amino groups of adenosine and cytidine are reasonable binding sites when deprotonated, and X-ray evidence is accumulating to support solution studies identifying these sites. The amino group of guanosine is expected to be an unfavorable site since N-1 deprotonation is favored. The exocyclic oxogroups are the least well-established binding sites. Cytosine derivatives most clearly can bind through the exocyclic oxygen, and good evidence is available to support this site in both solution and in the solid. In fact, O-2 may be the most important exocyclic site at neutral pH. Binding to the exocyclic oxygens of the guanosine, inosine, uridine, and thymidine will probably occur, if weakly, in the solid, but no noncontroversial solution evidence exists for participation of these exocyclic oxygens in metal binding.

The question of chelation involving an endocyclic nitrogen atom and exocyclic group has always been both intriguing and controversial. The N-3 strong, O-2 weak chelation mode of cytosine is well established. No truly irrefutable evidence for chelation of *purine* nucleosides has been presented. Some formulations of solid compounds have been interpreted to suggest such chelation.

Without a doubt, the cis hydroxyls of the ribose sugar can form chelates. This binding mode is best known for Os(VI) and Cu(II). In Cu(II) complexes of this type, the remaining problem is the nature of the remaining groups in the coordination sphere. Since the complexes are difficult to crystallize, and no X-ray structures have been reported. The Os(VI) complexes are very well characterized.

The nature of the binding to the phosphate groups of nucleotides is presently the biggest unsolved problem and presents a large number of unanswered questions. Are the interactions mostly indirect in the form of outer-sphere complexes? Information to answer this question is insufficient both for solution (where spectroscopic criteria are inadequate) and for the solid state (where polymerization clouds the issue and where formation of crystalline materials is a big impediment). Are "macrochelates" possible in which the base and the phosphate group both bind to the same metal? Solution studies indicate such chelates are possible, but no solid-state evidence is available.

Considerable progress has been made in understanding and confirming the nature of metal binding to the bases. The primary binding sites originally identified by Simpson and Eichhorn have in large measure been confirmed with newer and more powerful spectroscopic techniques such as NMR relaxation studies, ^{13}C-NMR, Raman and Raman difference spectroscopy, and X-ray crystallography. It has been possible in the case of cytidine to further define the nature of N-3 and O-2 binding in cytosine. One of the most important results that has emerged from the X-ray structural studies and not foreseen in early work is the involvement of the exocyclic groups in interligand hydrogen bonding interactions. A second important concept which is emerging is the steric requirement of the metal centers in binding to the endocyclic groups. The exocyclic groups may interfere with complex formation. Therefore, it is not surprising that the metal centers that have the greatest affinity for nucleic acid bases [Pt(II), Pd(II), Cu(II), Hg(II), and Ag(II)] also form complexes of *low* coordination number.

In this section we have briefly summarized both what is known and what is unknown about structures of small complexes. Despite the challenge ahead in establishing and answering the questions on small molecules—questions which have not yet been answered by the extensive recent activity in this area—even greater uncertainties are associated with the nature of metal binding to polymers.

2.6 Summary of Major Characteristic Structural Changes Induced by Metal Binding

A number of highly accurate structural analyses now exist and provide a basis for the evaluation of the effects of metal binding on the geometric parameters in the purine and pyrimidine bases. In general (5, 11), it appears that metal binding results in substantially smaller perturbations on the geometric parameters in the bases than protonation or alkylation at the same site that metallation takes place.

In an attempt to show some quantitative aspects of this general statement, we will examine briefly the case of metal binding at N-3 of cytosine derivatives for which there are accurate data on neutral, N-3-protonated, N-3-alkylated, and N-3-metallated species (Table 4). It is well known (123) that while protonation or alkylation at N-3 of cytosine causes minimal changes in bond lengths, there is a relatively large perturbation of the endocyclic bond angle at N-3, C-2–N-3–C-4. For example, the endocyclic bond angle at N-3 in 1-methylcytosine (124) is 120.0(1)°, and this angle increases by 4 to 5° on protonation [124.7(3)° in protonated 1-methylcytosine perchlorate (125)] or alkylation[123.7(8)° in 3-methylcytidine methosulfate · H_2O (126)]. In Table 4 we summarize the effect on the endocyclic bond angle at N-3 due to metal binding [Cu(II), Co(II), Pd(II), Pt(II), Ag(I), and Hg(II)] at N-3. The average value of the C-2–N-3–C-4 angle in the metallated cytosine derivatives at 120.6(10)° is

Table 4 Some Pyrimidine-Ring Parameters in a Variety of Neutral, N-3-Protonated, N-3-Alkylated, and N-3-Metallated Cytosine Derivatives

Compound	C-2–N-3–C-4 (degrees)	C-2–O-2 (Å)	C-4–N-4 (Å)	Reference
1. 1-Methylcytosine	120.0(1)	1.234(2)	1.336(2)	124
2. Protonated 1-methylcytosine perchlorate	124.7(3)	1.227(4)	1.318(5)	125
3. 3-Methylcytidine methosulfate·H_2O	123.7(8)	1.20(1)	1.31(1)	126
4. [Cu^{II}(Mesal)(cytosine)]NO_3	120.8(4)	1.237(6)	1.326(6)	127
5. Cu^{II}(glygly)(cytosine)	121.2(4)	1.234(6)	1.317(6)	128
6. Co^{II}(5′-CMP)(H_2O)	121(1)	1.23(2)	1.34(2)	129
7. $Pd^{II}Cl_2$(1-methylcytosine)$_2$	121.4(2)	1.230(2)	1.330(3)	94
8. Pt^{II}(bis(isopropyl)sulfoxide)-(1-methylcytosine)	120(1)	1.23(2)	1.33(2)	93
9. $Pt^{II}Cl_2$(DMSO)(cytidine)	119(1)	1.24(2)	1.35(2)	92
10. Ag^{I}(1-methylcytosine)(NO_3)	119.7(2)	1.253(3)	1.350(4)	97
11. $Hg^{II}Cl_2$(1-methylcytosine)	122(1)	1.23(1)	1.33(1)	91
Average (4 through 11)	120.6(10)	1.236(8)	1.334(12)	

virtually identical with that in 1-methylcytosine (124), 120.0(1)°. Similar data for the exocyclic bond lengths C-2–O-2 and C-4–N-4 again suggest that metallation at N-3 does not significantly perturb these parameters. The lone exception is in the complex Ag^{I}(1-methylcytosine)(NO_3) (97) where the C-2–O-2 bond length at 1.253(3)° Å is significantly longer than the observed value of 1.234(2)° Å in 1-methylcytosine (Table 4). The elongation in the C-2–O-2 bond length in this Ag(I) complex is undoubtedly due to the multiplicity and strength of the Ag–O-2 interactions that appear in this unusual structure (97), see Section 3.

More susceptible to intracomplex interactions are the exocyclic bond angles at N-3 (Table 5). As noted in Section 2.3, Cu(II) and Hg(II) are normally bound to cytosine derivatives by an N-3, O-2 chelation mode. where the interaction with O-2 is considerably less than with N-3. Nonetheless, the perturbations in the exocyclic bond angles at N-3 are substantial with the angle M–N-3–C-2 being on the average about 15 to 20° smaller than the M–N-3–C-4 angle (Table 5). As mentioned in Section 2.3, Pd(II) and Pt(II) interactions with O-2 are less significant, and the dissymmetry in the exocyclic bond angles at N-3-bound Pd(II) or Pt(II) complexes falls to about 6 to 10° (Table 5).

Table 5 Parameters for the M–N-3, and M–O-2 Intramolecular Interactions in a Number of N-3–Metalated Cytosine Derivatives

Compound	M–N-3 (Å)	M–O-2 (Å)	M–N-3–C-2 (degrees)	M–N-3–C-4 (degrees)	Reference
1. [Cu^{II}(Mesal)(cytosine)]NO$_3$	2.008(1)	2.772(1)	108.4(2)	130.6(2)	127
2. Cu^{II}(glygly)cytosine)	1.979(3)	2.819(3)	110.3(3)	127.8(3)	128
3. Pd^{II}Cl$_2$(1-methylcytosine)	2.031(2)	3.015(2)	116.0(1)	122.7(1)	94
4. Pt^{II}[bis(isopropyl) sulfoxide](1-methylcytosine)	2.058(7)	3.004(7)	115.3(7)	125.1(8)	93
5. Pt^{II}Cl$_2$(DMSO)(cytidine)	2.03(1)	3.06(1)	117(1)	124(1)	92
6. Hg^{II}Cl$_2$(1-methylcytosine)	2.17(1)	2.84(1)	107(1)	131(1)	91

Similar dissymmetry is observed in the exocyclic bond angles at N-7 in M–N-7-bonded 6-amino- and 6-oxo(thio)purines, owing to interligand hydrogen bonding (10), and to M–O-6 (weak) (10) or M–S-6 (strong) interactions. As noted in Section 2.1, N-7-bonded adenosine derivatives often show interligand hydrogen bonding involving the exocyclic amino group N-6 H_2. Some examples of the parameters in such interligand hydrogen-bonded complexes are presented in Table 6. In all cases the presence of the interligand hydrogen bond causes a striking dissimilarity in the exocyclic angles at N-7, with M–N-7–C-5 being about 20° larger than M–N-7–C-8, presumably owing to the steric requirements involved in the formation of the interligand hydrogen bond scheme (10). Similar adjustments are noted in interligand hydrogen-bonded 6-oxopurine complexes (Table 7). Weak M–O-6 interaction, in place of a hydrogen bonding mode, results in a reversal in this trend; and in such systems the M–N-7–C-5 angle is about 20° *smaller* than the M–N-7–C-8 angle (Table 7). This reversal is even more dramatic in the strongly bonded M–N-7, S-6 chelates found by Lippard (83) and Sletten (84) (Table 7).

Thus, while the bond lengths and endocyclic bond angles are only minimally perturbed by metal binding to purine and pyrimidine bases, the exocyclic bond angles at the site of metal attack will significantly adjust to take advantage of favorable interactions such as interligand hydrogen bonding or chelation. Unfortunately, it is not possible at this time to utilize these small structural differences in understanding the nature of the changes in NMR or vibrational spectra which accompany complex formation.

3 STRUCTURAL STUDIES MOST RELEVANT TO POLYMER BINDING

We turn now to the relationship between the structure of metal complexes of small molecules to the structure of metal complexes of polymers. The juxtapositions of the bases, sugar, and phosphate groups in polymers offer a considerable number of additional binding possibilities, and it becomes likely that such more complex structure can lead to metal interactions that are weak or nonexistent in monomer species.

Recently, Klug and co-workers (140) have been studying by isomorphous X-ray diffraction methods the binding sites for a variety of metal complexes and metal ions to one structural modification of yeast tRNA[Phe] (141). They found many types of metal binding sites: Mn(II), Co(II), and Os(VI)O_3pyr$_2$ bind to N-7 of a guanine base with other ligands in the primary coordination sphere (e.g., water), forming hydrogen bonds to neighboring oxygen or nitrogen atoms (as found in many monomeric

Table 6 Conformational Features About N-7 in Several Cu(II)–N-7-Bonded 9-Methyladenine Complexes

Complex	Cu–N-7 (Å)	Cu–N-7–C-5 (degrees)	Cu–N-7–C-8 (degrees)	Intramolecular Hydrogen Bonding	Reference
1. $Cu^{II}(Mesal)(H_2O)(9\text{-methyladenine})$	2.037(2)	136.2(2)	119.6(2)	$O(Mesal)\cdots H_2N\text{-}6$	130
2. $Cu^{II}(glyglы)(H_2O)(9\text{-methyladenine})$	2.021(4)	137.7(2)	116.2(2)	$H_2O\cdots H_2N\text{-}6$	25
3. $Cu^{II}(SO_4)(H_2O)_4(9\text{-methyladenine})$	1.995(2)	133.6(1)	120.5(1)	$O_3SO\cdots H_2N\text{-}6$	131
4. $[Cu^{II}(H_2O)_4(9\text{-methyladenine})]^{2+}$	2.008(2)	135.0(1)	120.4(1)	$H_2O\cdots H_2N\text{-}6$	132

Table 7 Conformational Features About N-7 in a Number of N-7-Bonded 6-Oxo- and 6-Thiopurines

Complex	M–N-7 (Å)	M–N-7–C-5 (degrees)	M–N-7–C-8 (degrees)	Intramolecular Interaction	Reference
1. Cu^{II}(Mesal)(theophyllinato)	1.986(1)	138.0(1)	118.9(1)	O-6 ··· HN(Mesal)	45
2. Cu^{II}(Mebenzo)(H_2O)(theophyllinato)	2.000(3)	136.1(2)	117.0(2)	O-6···HN(Mebenzo) O-6···H_2O	75
3. Cu^{II}(Me_2sal)(theophyllinato)	1.969(7)	127.5(4)	128.6(4)	None	133
4. Cu^{II}(Me_2benzo)(theophyllinato)	1.956(3)	117.8(2)	138.9(2)	Cu–O-6 = 2.919(3) Å	76
5. $Hg^{II}Cl_2$(guanosine)	2.16(2)	118(2)	134(2)	Hg–O-6 = 3.08(2) Å	134
6. Pd^{II}(6-thio-9-benzylpurine)$_2$	2.06(2)	109(1)	145(1)	Pd–S-6 = 2.31(1) Å	83
7. $Cu^{II}Cl_2$(6-thio-9-methylpurine)	1.992(1)	113.3(1)	142.0(1)	Cu–S-6 = 2.424(1) Å	84

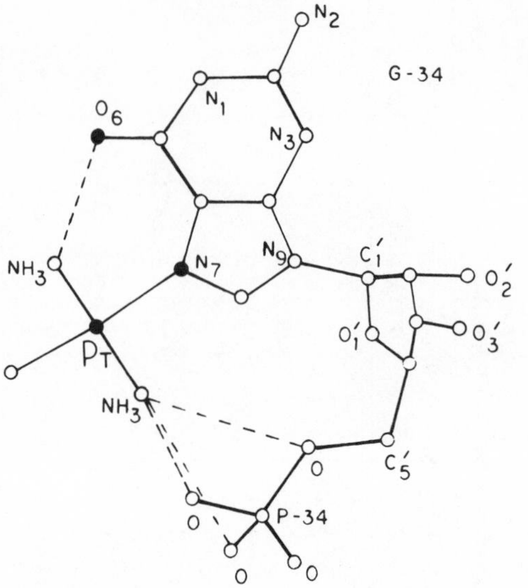

Figure 24 Proposed binding of *trans*-[Pt(II)Cl$_2$(NH$_3$)$_2$] to residue Gm34 (a l-methylguanosine site) of yeast tRNA^{Phe}. (After ref. 135.)

studies, Sections 2.1 to 2.3); Hg(II) binds to O-4 of an exposed uracil base (as found in the HgCl$_2$ complexes (109) of uracil and dihydrouracil, Section 2.4); Mn(II), Sm(III), and Lu(III) generally prefer regions rich in phosphate groups (some of which are coincident with Mg(II) binding sites in native tRNA), as expected for these hard metal ions.

We wish to concentrate here however on the binding of Pt^{II}Cl$_2$(NH$_3$)$_2$ to tRNA^{Phe} (140). As mentioned in Sections 1 and 2.2, reagents such as Pt^{II}Cl$_2$(NH$_3$)$_2$ have well-known antineoplastic activity as their cis isomers, while their trans isomers are inactive. Unfortunately, under the conditions employed by Klug (140), no binding was found for the cis isomer of Pt^{II}Cl$_2$(NH$_3$)$_2$, but the trans isomer shows a clearly defined binding site. For the trans isomer, the Pt binds directly to N-7 of a 1-methylguanosine residue (Gm34) at a distance of about 2.2 Å (Figure 24). Most interestingly, the trans ammine ligands participate in interligand hydrogen bonding to O-6 and the phosphate group of the same Gm34 residue (Figure 24), calling attention again to the important nature of such interactions. We (10) and others (5, 37) have suggested that the interligand interactions common in monomeric systems would also exist in complexes of polymers.

While the studies of Klug (140) do not provide any information about the binding of *cis*-[Pt^{II}Cl$_2$(NH$_3$)$_2$] to polymer, some recent studies on monomers allow interesting speculation on one increasingly mentioned mode of binding for the cis isomer, namely, intrastrand crosslinking (142). As mentioned briefly in Section 2.2, structurally similar Pt(II) complexes of 5′-IMP can be formed by the reaction of cis-$[Pt^{II}(NH_3)_2(H_2O)_2]^{2+}$ and cis-$[Pt^{II}(en)(H_2O)_2]^{2+}$ cations with the disodium salt of 5′-IMP. In each case, both water ligands are displaced by N-7 of two symmetry-related 5′-IMP ligands (Fig. 12 and Section 2.2). Moreover, the two complexes are both nonstoichiometric and structurally different from their analogous guanosine complexes (Section 2.2).

We (143) felt that it was important to try to understand the underlying reasons for these differences and have determined the structure of a Cu(II) complex of 5′-IMP. We believe the structure of this complex bears directly on the nonstoichiometry of the Pt(II)–5′-IMP complexes and also addresses the question of intrastrand cross linking of polynucleotides by Pt(II) and Cu(II). Reaction of $[Cu(II)(dien)]^{2+}$ (where dien = diethylene-triamine)] with Na_2(5′-IMP) produces the complex anion $[Cu(dien)(5′-IMP)_2]^{2-}$, which crystallizes as the hydrated monosodium salt (143). This Cu(II) complex, unlike the Pt(II) systems, is stoichiometric. The structure of this complex is very similar to but not isostructural with the Pt(II) salts. In fact, while the Cu(II) complex is stoichiometric, it is disordered, with the Cu(II) alternatively forming a strong equatorial bond to N-7 of one 5′-IMP ligand and a very weak axial bond with N-7 of a two-fold related 5′-IMP dianion. In an attempt to correlate these results, we (143) noted, as have others (71, 72), that the structures of the Pt(II) salts bear a strong resemblance to the structure of the monosodium salt of 5′-IMP. In the structure of the monosodium salt, a water molecule lies on a crystallo-graphic two-fold axis and links, via hydrogen bonds, two symmetry-related 5′-IMP dianions. In the isomorphous Pt(II) salts, this water site is partially occupied by the Pt(II) moiety, and the hydrogen bond scheme present in the monosodium salt is replaced by two cis coordination bonds. The $[Cu(dien)]^{2+}$ complex is disordered because of its inability in the presence of the tridentate dien chelate to form equatorial bonds to both of the two-fold related 5′-IMP ligands (143). In Figure 25 we attempt to show the relationship among the structures of the monosodium salt, the Pt(II) complexes, and the Cu(II)(dien) system.

In terms of the possible effect on a polynucleotide structure owing to intrastrand crosslinking, the structural similarities displayed by these 5′-IMP salts are particularly important. Our analysis (143) suggests that there is a strong competition between the crystal packing forces operative in the solid and the distortion of the basic structure due to the simultane-ous binding of the two 5′-IMP bases by the Pt(II) moiety. In fact, if one assumes that the crystal packing forces operative in the 5′-IMP complexes

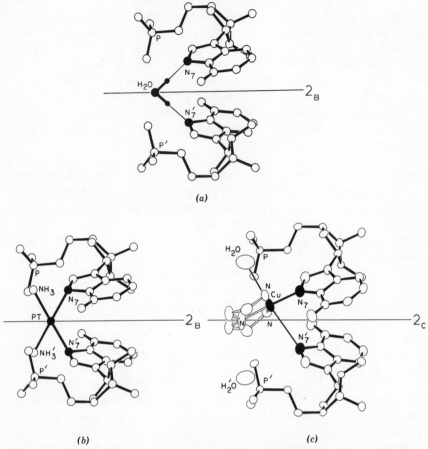

Figure 25 Comparison of the coupling of two-fold-related 5'-IMPs via (a) a water molecule in the structure of monosodium 5'-IMP (after ref. 69), (b) a diamminePt(II) moiety in the structure of [bis(5'-IMP)(NH₃)₂Pt(II)]²⁻ (after ref. 71), and (c) an aquodiethylenetriammineCu(II) moiety in the structure of [bis(5'-IMP)(aquo)(dien)copper(II)]²⁻. (After ref. 138.)

approximate those found in a base-stacked polynucleotide, the above results suggest that intrastrand crosslinking by reagents such as *cis*-[Pt(NH₃)₂Cl₂] and *cis*-[Pt(en)Cl₂] will place a significant strain on the conformational properties of a polynucleotide. For Cu(II) complexes, the demands imposed on the polynucleotide framework will be minimized owing to the possibility of an equatorial-axial mode of intrastrand crosslinking.

In another current area of investigation, platinum(II) complexes containing planar aromatic ligands (15) appear to bind to double-stranded

DNA through an intercalative mode. Intercalation of planar, aromatic, cationic heterocycles such as ethidium and acridine dyes is an established mode of interaction in nucleic acid chemistry. Lippard (15) has been able to determine (through chemical and biochemical assays) that the cationic Pt(II) intercalators bind competitively with the nonmetallo intercalators and therefore probably bind in essentially the same fashion and at the same sites. Most importantly, Lippard, Bond, Wu, and Bauer (144) have investigated the fiber diffraction patterns of intercalatively bound Pt(II) reagents with duplex DNA. The electron-dense Pt greatly facilitates the analysis. The mutual exclusion model (simply stated, only every other possible site between stacked bases in a duplex polynucleotide is available for intercalative binding) accounts for the diffraction data for the Pt(II) intercalators. The competitive binding between the metallo intercalators and ethidium (15) also suggests that the mutual exclusion model is appropriate for aromatic drug–DNA binding.

A recent structural investigation (97) of the complex formed between AgNO$_3$ and 1-methylcytosine provides some insights into the binding of Ag(I) to polynucleotides. The interaction of Ag(I) with nucleic acids has been widely studied (1, 2, 145, 146) and occurs primarily at guanosine–cytidine (G–C) regions of DNA. Although this preferential binding has been exploited to separate nucleic acids of different G–C content (146), there were until this work (97) no structural studies on Ag–cytidine complexes, and there remain to our knowledge no structural studies on Ag–guanosine complexes.

Some aspects of the structure of the [(1-methylcytosine)Ag^I]nitrate complex are illustrated in Figure 26. The most pronounced feature is the formation of centrosymmetrical dimers in which the 1-methylcytosine ligands bridge two symmetry-related Ag$^+$ ions. Within these dimers, there are two strong metal–ligand bonds—one to N-3 [2.225(2) Å] and one to O-2 [2.367(2) Å]. Interestingly, these dimeric units are formed into columns and connected by a second Ag–O-2 bond at 2.564(3) Å. Within these columns there is significant base–base overlap (mean distance 3.34 Å). The role that O-2 plays, both within the dimeric units and the propagation of the dimers into a column, is especially interesting. Moreover, the Ag $\cdot$ $\cdot$ $\cdot$ Ag repeat length (3.64 Å) in the polymeric columnar stacks is reminiscent of the base–base stacking distance of about 3.5 Å in duplex DNA (147). The nearly commensurate Ag $\cdot$ $\cdot$ $\cdot$ Ag and base–base distances suggest that cooperative propagation of base–Ag–base polymers parallel to the helix axis of a duplex DNA could be induced. In light of the versatility of O-2 of cytosine residues, such polymeric fragments could be readily accommodated in regions of high G–C content (97).

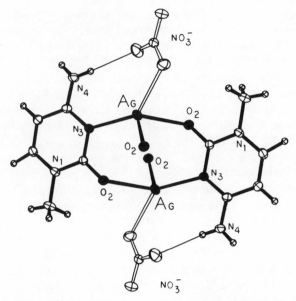

Figure 26 Basic dimeric unit formed in (nitrato)(l-methylcytosine)silver(l). (After ref. 97.)

4   EFFECTS OF METALS ON POLYNUCLEOTIDE STRUCTURE

4.1   Structures of the Polymers

Nucleotides are the monomeric building blocks of polynucleotides or nucleic acids, which can be thought of as arising from the condensation of the 5′-phosphates with the 3′-hydroxide groups of 5′-mononucleotides. A polynucleotide therefore contains a backbone of ribose or deoxyribose alternating with phosphate groups, each of which is bound to the 3′-OH of one sugar and the 5′-OH of the other sugar. Attached to the sugar portion of this backbone are the nucleotide bases. Monoribonucleotides of course lead to polyribonucleotides, or RNA, and monodeoxynucleotides lead to polydeoxynucleotides, or DNA. The primary structure of polyribonucleotide is shown in Figure 27.

Polynucleotides can exist as unordered structures, or random coils, or they can be organized into helices. The latter may be simple helices involving the ordering of individual polymer strands, or they may be multiple helices, in which two or more polynucleotide strands are intertwined. The best-known multiple helix is the DNA duplex, in which each base from one strand is hydrogen bonded to a base on the other strand; the resulting base pairs are parallel to each other and perpendicu-

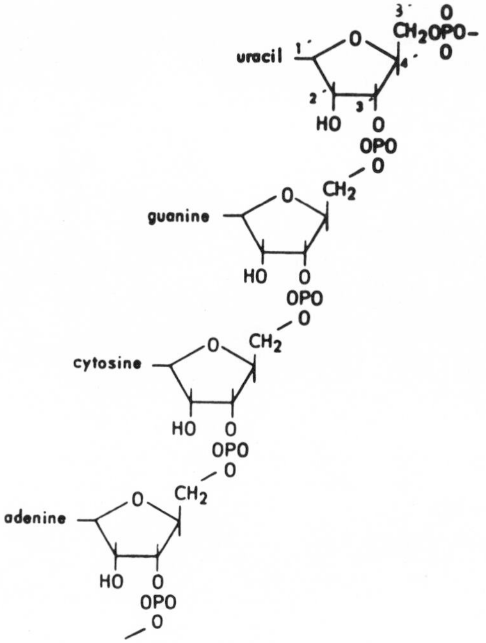

Figure 27 Primary structure of polyribonucleotide backbone.

lar to the helical axis. The base pairs are either guanine–cytosine (G–C) or adenine–thymine (A–T) (Figure 28).

4.2 Metal Binding to Polynucleotides

The same functional groups that are available for metal binding in mononucleotides, namely, phosphates, base O and N atoms, and sugar hydroxyls, are of course also available in the polymers, and one can therefore expect similar bond types. We will first discuss some evidence that the same binding sites that are involved in metal binding to the base in monomers are also found in polymers. However, we emphasize that relatively little definitive information is available and furthermore that spectroscopic techniques are most useful in detecting base binding.

Some stoichiometric information is consistent with equivalent binding in monomers and in polymers. An excellent example is the work of Ottensmeyer and Whiting (135), who found that the complex $Pt^{II}(DMSO)Cl_3^-$ formed 2:1 Pt:Ade complexes with poly(A) and 1:1 complexes with poly(C), poly(G), and poly(U). At higher pH, they obtained evidence for further reaction of Gua moieties, consistent with a

A

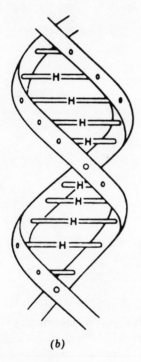

Adenine   Thymine

(b)

Figure 28 (a) The adenine–thymine base pair, showing two hydrogen bonds. Similarly, guanine and cytosine share three hydrogen bonds. (b) Schematic representation of DNA double helix. Ribbons indicate polyribose phosphate backbone. Parallel bars represent hydrogen-bonded bases.

second Pt adding to N-1. For DNA, the stoichiometry that can be predicted from our knowledge of monomer binding was also found. In another study, Chang, Beer, and Marzilli (136) observed that the reagent OsO_4/bipyridine gave stoichiometries with polynucleotides that were exactly as predicted from Behrman's monomer studies (4, 137). In addition, it was possible to evaluate the electronic spectrum of the polymer complex and to demonstrate that it was consistent with the structure known for monomers.

Raman spectroscopy is at present a very promising technique for investigating metal binding to polymers. The extensive investigations by

Tobias have provided a "library" of spectral changes which can be used for identifying metal binding sites. This approach is most useful for observing base binding, and results to date suggest the same sites are involved in polymers as are involved in monomers (39).

NMR spectroscopy is more difficult for polymers, and generally the polymer must be denatured in order to observe sharp resonances. NMR line broadening studies furnish evidence for such similarity in complexes of copper(II). The latter preferentially broadens H-8, rather than H-2, of 3'-AMP and 5'-AMP as well as of poly(A), the polyribonucleotide containing only adenine bases (46). Since H-8 broadening in the monomers is correlated with N-7 binding to metals (46, 48), it appears that Cu(II) also binds to N-7 in poly(A). Similarly, Cu(II) binding to N-3 in poly(C) as well as in 5'-CMP, and in poly(U) as well as in 5'-UMP, are indicated (138). Such studies implicate only N-7 in poly(I), however (138), even though both N-1 and N-7 binding were demonstrated in 5'-IMP (138) and inosine (148). Since N-1 in poly(I) is engaged in hydrogen bonding to form a quadruple helix, only the N-7 group is available for Cu(II) complexing without disruption of the ordered poly(I) structure.

Proton relaxation measurements on Mn(II) complexes of poly(A) have suggested that Mn(II) also binds preferentially to N-7 in the five-membered ring compared to binding to sites on the six-membered ring. In fact, in solutions containing a large excess of the polynucleotide ligand, 39% of the Mn(II) was supposedly bound to N-7, with only 13% bound to either N-1 or N-3. Perhaps the most important conclusion from these studies is that 100% of the Mn(II) is bound to two phosphate groups. Such solutions therefore contain at least three types of complexes: (1) Mn(II) bound to two phosphates, (2) Mn(II) bound to two phosphates and N-7, and (3) Mn(II) bound to two phosphates and a six-membered ring nitrogen site (149).

This study demonstrates clearly that a definition of the metal binding sites on the liganding polymers does not tell the whole story about the resulting structures of the polymer complexes, since a variety of combinations are possible. The effects of metal ions on the secondary and tertiary structure of the polymer are of as much importance as the sites of coordination.

Although precise binding modes of metals to polynucleotides have generally not been determined, the relative affinity of metal ions for G–C or A–T rich regions of DNA have been established and are in fact notably different. Thus Cu(II) (150), Hg(II) (151), and platinum (152) prefer G–C rich regions while Ag(I) ions prefer A–T (153). The dependence of Hg(II) binding to DNA upon the base composition of DNA has been successfully applied to the separation of DNAs of different base contents (154).

When rare earth ions are bound to DNA, they engage in energy transfer resulting in fluorescence. The fluorescence of terbium(III) DNA complexes has been used as a sensitive probe for DNA in chromatin (155).

4.3 Conformation Changes of Polynucleotides Produced by Metal Ions

Polynucleotides may be randomly oriented in space in a so-called "random coil structure," or they may be organized into ordered conformations. The latter are stabilized by two types of interactions: (1) stacking of the heterocyclic bases by interactions of the π electrons on the bases, and (2) hydrogen bonding. Ordered structures may be simple helical, in which the polynucleotide strands do not interact, or multiple helical conformations that involve hydrogen bonding between bases on different strands.

The effect of metal ions upon polynucleotide structure depends in part on their relative ability to bind phosphate or bases. Phosphate binding tends to stabilize ordered structures, while base binding tends to weaken such structures. For example, poly(A)–polyriboadenylic acid exists as a single helix at pH 7 and as a double helix at pH 5. Base-binding Cu^{2+} ions eliminate the ORD curve characteristic of poly(A) single helix at pH 7 and of poly(A) double helix at pH 5 (156). Phosphate-binding Ni^{2+} and Co^{2+} ions stabilize the single helix at the expense of the double helix; the latter involves hydrogen bonds between phosphate oxygens and amino nitrogens, and these metal ions thus disrupt the double-helical conformation by competing for the phosphate groups (156). It is in fact possible to regulate the proportion of random-coil, single-helix, and multiple-helical forms of polynucleotides generally by the presence of different divalent metal ions modulated by ionic strength and pH (156, 157).

Contrary to the generalization made above, base-binding metal ions are sometimes capable of promoting formation of an ordered structure. Ag^+ ions, which bind strongly to poly(U) and to poly(I), do indeed induce helical structures in both polynucleotides, although neither has a strong tendency to helix formation. In fact, poly(U) generally is unordered except at very low temperatures, and poly(I) generally requires high ionic strength for its multiple helical orientation. Ag^+ ions produce ordered forms by crosslinking the bases of two polynucleotide strands, and in effect produce structures similar to hydrogen-bonded base pairs, with Ag^+ ions replacing the hydrogens (158). Thus the generalization that base-binding metal ions interfere with hydrogen-bonded helices is not disobeyed, since the Ag^+-stabilized helices are not hydrogen-bonded.

Curiously the same metal ions (e.g., Cu^{2+}) which readily disintegrate ordered, or stacked, polynucleotide structures are in fact capable of stacking mononucleotides (159). It is not difficult to see why Cu^{2+} ions can have such opposite effects on mono- and polynucleotides. Mononucleotides of course are not bound to each other before metal binding, so that the metals can hold the nucleotides together in such a way as to produce stacking. The same metal ions binding to bases in polynucleotides will seek to order the nucleotides in a manner dictated by the stereochemistry of the metals, which may be different from the way in which the phosphodiester bonds orient the bases with respect to each other. Metal–base binding in the polymers therefore results in counteracting the constraints imposed by the polymeric linkages and therefore causes the disordering to take place. Metal–base binding in mononucleotides can induce an ordering of bases which is not opposed by internucleotide linkages.

In addition to their ability to form bonds to the heterocyclic bases and to the phosphate groups of polynucleotides, metal ions can intercalate between the stacked bases by intercalating with the π electrons of two adjacent bases on a polymer chain. In this way the helix is distorted by increasing the distance between the intercalated stacked bases. Intercalation has been proposed to account for intermediates in the denaturation of DNA by zinc (159) and has been demonstrated most effectively for the reaction of DNA with a platinum(II) complex of terpyridine by Jennette et al. (161).

4.4 Crosslinking, Unwinding, and Rewinding

4.4.1 Crosslinking. As has already been noted, metal ion interactions with nucleic acids can involve the formation of crosslinks between nucleotides. Crosslinks involving Cu^{2+} ions have received particular attention (162). The disordering of poly(A) and poly(C) by Cu^{2+} ions was monitored spectrophotometrically, and it was found that the disordering was dependent on the concentration of the polymer. This concentration dependence indicated that the Cu^{2+} ions interact with more than one polymer strand and therefore form crosslinks between the strands (162).

It was also found that the disordering of the polynucleotides by Cu(II) is very cooperative; that is, the binding of one copper ion to the polymer increases its affinity to other Cu^{2+} ions, and each Cu^{2+} ion causes the disordering of 4-5 bases. In spite of the fact that this "cooperative unit" is only 4-5 bases long, the disordering by a concentration of Cu^{2+} ions of poly(A) is much greater than that of hexaadenylic acid. The discrepancy can be explained by the participation of intramolecular crosslinks in poly(A) and the absence of such crosslinks in the hexamer, which can accommodate only intermolecular crosslinks. Thus both intermolecular and intramolecular crosslinks by Cu^{2+} ions have been demonstrated. Though the crosslinks must involve at least one bond to a base—metals such as Mg(II) that bond only to phosphate do not produce these effects—it is not clear whether both bonds are to bases or whether one bond is to phosphate and the other to base (162).

Mg(II), which bonds only to the phosphate groups of polynucleotides, incidentally involves completely anticooperative binding (163).

4.4.2 Unwinding and Rewinding.

The formation of crosslinks between DNA strands by metal ions makes possible the reversible unwinding and rewinding of DNA (160, 164–168). In the absence of divalent ions, heating native double-stranded DNA generally results in denaturation or in unwinding of the double helix. When the denatured DNA is cooled, there is a tendency to reform hydrogen bonds, but there is a constraint that requires G to bond with C and A to bond with T. Since the single DNA strands become random coils during the denaturing process, the complementary bases are no longer in register and therefore are prevented from reforming the double helix (164).

By forming crosslinks between the DNA strands during the unwinding process, base-binding divalent metal ions can maintain the complementary bases in close proximity, so that when the thermodynamic conditions for stability of the double helix return, the bases are then in register and therefore they hydrogen-bond and reproduce the double helix. Thus heating DNA in the presence of Zn(II) causes the DNA to unwind, and then cooling it brings about rewinding, under conditions that would not permit rewinding in the absence of the Zn(II) (160).

The complete reversibility of the unwinding and rewinding of the DNA double helix merely by temperature manipulation appears to be practically unique for zinc ions, among the metals that have been studied. Presumably zinc confers just the right degree of stability to the bonds to the DNA to permit crosslinking as well as ready decomplexation in favor of hydrogen bonding at low temperature. Metals like Cu(II) that form stronger bonds with DNA also produce crosslinks with the unwound strands at elevated temperature, but these crosslinks are retained at low temperature and can only be dislodged by treatment with complexing agents (166) or strong electrolyte (164). A requirement for the metal ion to be able to promote reversible unwinding and rewinding of DNA is that the metal ion can form bonds with the nucleotide bases.

Metal ions such as Hg^{2+} that bond very strongly to the DNA bases can form a large number of very strong crosslinks to the DNA, so that the resulting DNA is not a random coil but rather a double helix with Hg bonds replacing H-bonds (169). The requirement for crosslinking is not met by methylmercury(II) ions, and these ions in fact react with DNA to denature it, and renaturation in the presence of CH_3Hg^+ is not possible (170, 171).

Polyribonucleotide double helices [poly(A) · poly(U) and poly(I) poly(C)] can also be unwound and rewound by metal ions (172).

4.5 Binding of Metal Ions to tRNA

The binding of metal ions to transfer RNA has received particular attention since the discovery that metal ions are required to stabilize the native amino acid-bearing form of the molecule (173, 173a). Evidence for a number of metal binding sites as well as more numerous weaker binding sites for metals on tRNA were obtained from solution studies (174–180). Mg^{2+} ions are particularly effective in stabilizing tRNA structure, but they can be replaced by other divalent ions, and the presence of univalent ions in high concentration can decrease the number of divalent ions required.

Recently these solution studies have received confirmation by X-ray diffraction studies that have pinpointed the sites of attachment of the metal ions to specific bases on yeast tRNA^{Phe}. Holbrook et al. (181) have identified the positions of Mg^{2+} ions in the structure from high electron density peaks, and have concluded that Mg^{2+} ions are particularly well suited for their stabilizing role because of the size and geometry of the hydrated Mg^{2+} ion, which apparently fits well into certain crevices in the tRNA structure. They have identified one Mg^{2+} bound to phosphate oxygen of residue 19, with a coordinated water molecule that is hydrogen bonded to guanine residue 20 (see Chapter 4). A second Mg^{2+} is very close to the first and bound to two phosphate oxygens on residues 20 and 21 (181). Another Mg^{2+} apparently fits into the right turn between residues 8–12 through 5–6 and hydrogen bonds to phosphate groups. A fourth Mg^{2+} ion is in the anticodon loop, bound to phosphate of residue 37, and is again attached to various bases in the anticodon loop through interligand hydrogen bonding. Depending upon the number of phosphate groups bound to these magnesiums, they are coordinated as $Mg(H_2O)_4^{2+}$, $Mg(H_2O)_5^{2+}$, or $Mg(H_2O)_6^{2+}$. Three of these sites have also been identified by Jack et al. (182), who have confirmed the existence of a large number (some 50) of weak Mg^{2+} binding sites. These workers (182) have also studied the binding of various other metals to tRNA. Rare earth ions (Sm^{3+} and Lu^{3+}) bind to five sites, three of which are identical to the Mg^{2+} binding site.

Trans-[Pt(NH$_3$)$_2$Cl$_2$] binds to N-7 of the methylguanine residue 34, with one NH_3 group hydrogen bonding to O-6 of that base and the other NH_3 group forming three H bonds to the phosphate group (Figure 24). A slight modification in the structure of the tRNA prevents the platinum from complexing, apparently because the structural requirements for binding five positions on the same nucleotide are very rigid. For the same reason cis-[Pt(NH$_3$)$_2$Cl$_2$] does not bind at all to tRNA (183).

Mercury(II) binds to O-4 of uracil-47 (182) in a manner similar to the Hg binding to uracil alone (109). Hg(II) also binds to the sulfhydryl group of the s^{4}U base in tRNA^{Val} (184).

Thus, in addition to its requirements for divalent metal ions such as Mg^{2+} for stabilization, tRNA binds to heavy metal ions in a manner that is not unpredictable from the reactions of these ions with monomers.

4.6 Catalysis of Degradation and Formation of Polynucleotides by Metal Ions

4.6.1 Degradation. The binding of metal ions to the phosphate groups of polyribonucleotides can result in the cleavage at elevated temperatures of the phosphodiester linkages and the consequent degradation of the polynucleotides (1, 185–188). This reaction occurs only with polyribonucleotides, not with the deoxynucleotides, and has been studied with a variety of divalent ions and rare earth ions. Pb(II) degrades most rapidly (186), while Zn(II) is the most effective degrading agent in the first transition series elements.

The fact that RNA, but not DNA, is degraded by metal ions indicates of course that the 2'-OH group is implicated in the reaction mechanism. The latter proceeds through the binding of the metal to the phosphate, which results in a positive dipole on the phosphorus. The negative charge on the 2'-OH is attracted by this positive dipole and forms a 2',3'-cyclic phosphate, thus displacing the bond to the 5'-phosphate group, which is cleaved. Finally the cyclic phosphate is hydrolyzed, and either the 3'- or the 2'-phosphate remains (Figure 29) (190).

The low susceptibility of DNA to phosphodiester bond cleavage has been cited as a possible explanation for the choice of DNA, rather than RNA, as the primary genetic material (191).

The rate of cleavage of phosphodiester bonds by metal ions depends on the adjacent base (188, 192) as well as on the tertiary structure of the polymer (193). The importance of tertiary structure is clearly demonstrated by the limited cleavage by Pb(II) (in 1 M NaCl) of yeast tRNA^{Phe}, which occurred at only two sites. Though both sites happened to be adjacent to dihydrouridines, the susceptibility to cleavage at these bases was attributed to their position on the tRNA, since poly(hU) was cleaved less readily than poly(U) and the bond adjacent to uracil in the tRNA was not cleaved.

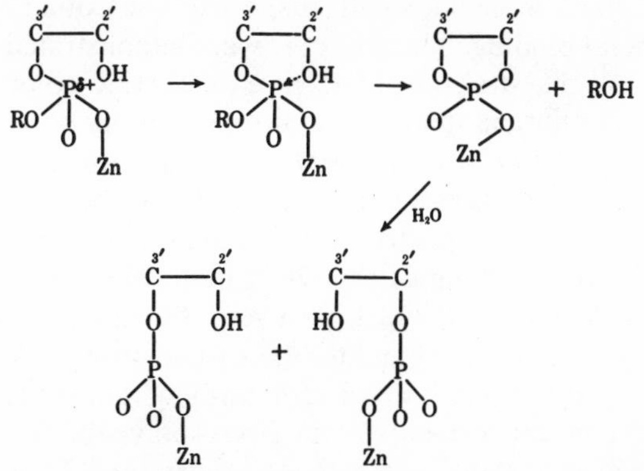

Figure 29 Mechanism of cleavage of phosphodiester bonds in RNA. [Reprinted with permission from *Biochemistry*, **10**, 2023 (1971). Copyright by the American Chemical Society.]

4.6.2 Formation of Oligonucleotide Bonds by Metal Ions. An activated derivative of 5'-AMP, adenosine-5'-phosphorimidazolide, which has an OH group from the phosphate replaced by an imidazole group, can be made to condense into oligonucleotides through catalysis by metal ions (194, 195). This process is of course a reversal of the degradation reaction. As in the degradation, Pb(II) is again the most effective catalyst (195), but the relative activity of other metals is not the same as for the degradation, so that Zn(II), for example, is less effective than Co(II).

4.6.3 Metal Ions in the Enzymatic Degradation of Nucleic Acids. There are a large number of enzymes whose function is the degradation of nucleic acids. Some of these require metal ion activation, and among these is bovine pancreatic deoxyribonuclease I.

The specificity of this enzyme, namely, the bases next to which cleavage occurs by action of the enzyme, can be markedly altered by binding metal ions to DNA before treatment with the enzyme. If the DNA is bound to a metal that preferentially binds to guanine [Cu(II)], the enzyme appears to recognize that this part of the DNA chain has been tampered with, and little cleavage then occurs at G sites. Similarly, if DNA is bound to a metal that preferentially binds to thymine [Hg(II)], cleavage at T sites is greatly diminished (196). Thus metals binding to DNA can produce qualitative changes in the mechanism of enzyme action.

4.7 Metal Ions Influencing the Packing of DNA Molecules

The reaction of metal ions with polynucleotides not only influences structure at the primary and secondary level. The packing, or the arrangement of DNA molecules with respect to each other, is also greatly affected by metal binding. This effect is well demonstrated by the impact of metal ions on the structure of DNA polypeptide complexes.

The positively charged quaternary nitrogen atoms on polylysine react with the negative charges on DNA phosphates to form a DNA–polylysine complex that is characterized by a very strong negative band in the circular dichroism (CD) spectrum. The intense dichroism is due to an anisotropic ordered packing of the DNA molecules. Some base-binding metal ions [Cu^{2+} and $Pt(NH_3)_2Cl_2$] convert the structure from one producing a strong negative CD band to one with a strong positive CD (197). Ag^+ ions, on the other hand, favor structures with a strong negative CD band and will convert a positive-CD DNA–polypeptide complex [e.g., DNA–poly($Lys_{50}Ala_{50}$)] into a negative-CD structure. Thus metal ions are very important in determining how DNA molecules are packed together.

5 BIOLOGICAL FUNCTION

The major reason for studying the reactions of metal ions with nucleic acids and their components and derivatives is of course to understand how metal ions are involved with the nucleic acids in biological processes. In fact, metal ions are essential to virtually all of these processes, but they can also generate deleterious effects. It is therefore very important to understand the conditions under which metal ions are beneficial or harmful.

5.1 Metal Ions and Chromatin

Chromatin is a complex of DNA and protein molecules in the nucleus of the cell. The DNA contains the genetic information, and the proteins are believed to be involved in the regulation of genetic expression. Chromatin contains DNA in a packed form. We have seen that metal ions can influence the packing of DNA complexed to polypeptides and that this influence operates in chromatin. The importance of metal ions in chromatin structure is dramatically illustrated by electron micrographs of the cell nucleus in the presence of high and low Mg^{2+} ion concentration (198). At high Mg^{2+} concentration dense particles of chromatin can be observed; these are diffused beyond recognition at lower concentrations. It is apparent that metal ion concentration is of significance in the organization of the chromosome.

The binding of the nuclear proteins to DNA is modified by enzymatic reactions that methylate, acetylate, and phosphorylate the proteins.

These modifications are believed to be involved in genetic expression. Metal ions modulate the phosphorylation reaction (199, 200) and therefore may be of importance in turning genes on and off.

5.2 Metal Ions and DNA Synthesis

The retention of genetic information by an organism when cells divide to produce more cells requires the replication of the DNA molecules. The replication process is an exceedingly complicated one that requires a number of DNA polymerases as well as other enzymes. Thus no single enzyme is alone responsible for replication in vivo, but enzymes, like *E. coli* DNA polymerase I, do catalyze the synthesis of DNA in vitro and are in fact part of the in vivo DNA synthesizing machinery. The enzymes require DNA templates, whose nucleotide sequences they copy, and deoxynucleoside triphosphates, which lose pyrophosphate during the elongation process. DNA polymerases generally require metal ions for their activity. We shall concern ourselves with *E. coli* DNA polymerase I, which has been by far the most extensively studied.

The enzyme contains two atoms of zinc(II) per molecule of protein (201). There is evidence that the zinc binds the growing end of nascent DNA to the enzyme (201). The catalytic activity also requires the participation of activating divalent metal ions, such as Mg^{2+} or Mn^{2+}, on other sites of the enzyme, and there is substantial evidence that these ions bind the nucleoside triphosphate substrates (202).

Sloan et al. (202) have obtained some rather exciting clues to the ways in which the activating metal ions may aid in DNA synthesis. Using proton and phosphorus relaxation studies on the adenosine and thymine deoxynucleoside triphosphates (dATP and dTTP) in the presence of Mn(II) with or without the enzyme, they have found that the distances between the Mn(II) and substrate protons and phosphorus atoms in enzyme–metal–substrate complexes are very different from those in the binary Mn(II) nucleotide complexes. Mn(II) binds to all three of the phosphoryl groups, α, β, and γ, in the nucleotide complex, but only the γ phosphoryl group is directly coordinated in the ternary complex (Figure 30). In the case of dTTP the change in bond distances between Mn(II) and the nucleotide produced by the enzyme leads to a change in the torsion angle about the glycosidic bond joining the deoxyribose to the base (202). In the binary complex this angle (ψ) is 40°, which means that the glycosidic bond has been rotated 40° from the position in which the N-1–C-6 bond in the pyrimidine is *cis* planar with the O-1′–C-1′ bond in the deoxyribose (203). In the ternary, enzyme-containing complex, the torsion angle becomes 90°. It has been noted (vide supra) that metal

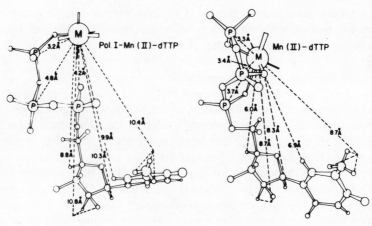

Figure 30 Diagrams of structures of Mn(II) complex of dTTP in the absence of DNA polymerase I (Pol I) (right) and Mn(II) complex with dTTP and Pol I (left). (Reprinted with permission from ref. 202.)

complexing can change the bond angles in nucleotides; this work demonstrates the change in bond angle produced in a metal-nucleotide complex through the binding of a second ligand (the enzyme) to the metal.

The 90° glycosidic bond angle achieved in the enzyme–metal–substrate complex is identical to the bond angle found in the usual B form of the DNA double helix. Thus it appears that one possible function of the enzyme, and of the metal ion, is to place the nucleotide into the proper conformation required for base pairing in the double helix. In this connection, it is observed that the binary complex Mn(II)–dATP already has a 90° glycosidic bond angle, and this bond angle is *not* changed as a consequence of binding to the enzyme (202).

The steric relationships found in the ternary complex may favor the removal of the pyrophosphate group through direct binding of Mn(II) to the γ (terminal) phosphate and the placing of a strain on the phosphate group. The structure is in fact admirably tailored to permit the formation of a phosphodiester bond along with pyrophosphate elimination (202).

It is possible that the placement of the glycosidic angle in the proper position for hydrogen bonding in the double helix helps to decrease the error in base selection during the formation of the DNA duplex. This error is in fact greatly dependent on the nature of the metal ion used to activate DNA polymerase (204–206) (see Chapter 3).

In addition to Mg(II) and Mn(II), it has been found that Co(II), Ni(II), and Zn(II) will activate a number of DNA polymerases (204). The enzymes with Mg(II), Mn(II), Co(II), and Ni(II) have been tested for fidelity by determining, for example, how much dCMP and dTMP is incorporated, using a poly[d(A–T)] template (205). For these studies a DNA polymerase from avian myeloblastosis virus was used, because this enzyme, unlike the *E. coli* enzyme, has no nuclease activity and therefore cannot remove errors. The incorporation of dTMP of course is a hit, and the incorporation of dCMP is a miss. Errors were produced with all of these metal ions, but Mg(II) and Ni(II) produced fewer errors (1/1400) than Co^{2+} (1/1100) and Mn^{2+} (1/600). It was postulated that the differences in fidelity in the presence of different metal ions may be due to the stabilization of different nucleotide conformations in the enzyme–metal–substrate complexes, so that the nucleotides would be more or less favorably aligned for double-helix formation.

Ni(II), as an activator of DNA polymerase, does not increase the error rate in DNA synthesis above that with Mg(II). However, when Ni(II) is added to the Mg(II)-activated enzyme, the error rate is greatly enhanced (205). A large number of metal ions have been tested to determine whether their addition to the DNA synthesis system would lead to a decrease in the fidelity of the enzyme. Of the Group 2A elements, for example, only Be(II) leads to such a decrease, while Ca(II), Sr(II), and Ba(II) produce no change in fidelity [although Ca(II) was inhibitory for both "correct" and "incorrect" incorporation] (207). The tendency of a metal to decrease fidelity in DNA synthesis could be correlated with carcinogenicity (208). In general, base-binding metal ions seem to induce a decrease in fidelity, whereas phosphate-binding metals have no such effect.

In addition to the errors involved in the incorporation of the wrong bases into DNA, metal ions also influence errors in the incorporation of the sugar moiety, that is, whether or not ribonucleotides are incorporated instead of deoxynucleotides. In the presence of Mg(II) only deoxynucleotides are utilized by the enzyme, but Mn(II) causes the error incorporation of ribonucleotides (209). This error incorporation has been employed by Salser et al. (210, 211) to synthesize DNAs that have only one of the four bases present as the ribonucleotides. Since the DNA synthesized by DNA polymerase is a copy of the template DNA, the modified poly(ribodeoxynucleotide) can then be used to determine DNA sequence (210, 211).

5.3 Metal Ions and RNA Synthesis

RNA constitutes the link between the DNA genes in the cell nucleus and the cytoplasm of the cell, where the genetic information is used to synthesize proteins. RNA molecules are synthesized on the DNA template

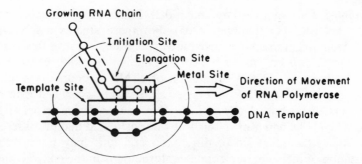

Figure 31 Scheme showing initiation and elongation sites in RNA polymerase. [Reprinted with permission from *Biochemistry*, **16**, 3331 (1977).] Copyright by the American Chemical Society.

under the influence of RNA polymerase and then move into the cytoplasm for protein synthesis. RNA polymerases can be found in all types of cells, but we shall confine our discussion to those from procaryotic cells and especially from *E. coli*.

The metal ion requirements of RNA polymerase are very similar to those of DNA polymerase. Two atoms of zinc are tightly bound to the enzyme (212). When the RNA polymerase is isolated from *E. coli* that has been grown in a medium containing Co(II) but deficient in Zn(II), the enzyme contains Co(II) instead of Zn(II) (213). The Co(II) enzyme has many properties similar to those of the zinc enzyme.

Also like DNA polymerase, RNA polymerase requires divalent metal ions at another site for activation. Mg^{2+} ions are the usual activators, but Mn^{2+} or Co^{2+} may be substituted (214–216).

The study of the activating metals to RNA polymerase is complicated by the fact that the active site of this enzyme consists of two components: (a) an initiation site that can be filled by a purine nucleotide or a dinucleoside monophosphate (217), and (b) an elongation site that binds any nucleoside triphosphate with a base complementary to one on the DNA template. RNA synthesis begins when an initiator binds at the initiation site. According to one mechanism of RNA synthesis, the enzyme moves along the DNA template, and at each base along the transcribing chain a substrate binds at the elongation site. This substrate becomes part of the growing RNA chain, eliminating pyrophosphate, forming a phosphodiester linkage, moving to the initiator site, and being released from that site (Figure 31).

The association of Mn(II) with the elongation site of the Mn(II) RNA polymerase has been demonstrated by Mildvan and co-workers (218, 219). In one study (219) the dinucleoside monophosphate ApU

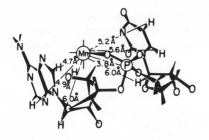

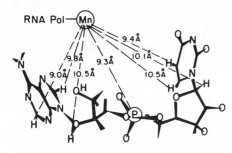

Figure 32 Diagrams of structures of Mn(II) complex of adenyl-(3'→5')-uridine in the absence of RNA polymerase (top) and in the presence of the enzyme (bottom). [Reprinted with permission from *Biochemistry,* **16,** 3332 (1977). Copyright by the American Chemical Society.]

[adenylyl-(3'→5')-uridine] was bound to the initiation site while ATP was bound to the elongation site. ApU binds only to the initiation site, thus precluding ATP binding there. Bond distances between Mn(II) and both ApU and ATP were determined in the same way as in DNA polymerase both in the presence and absence of the enzyme. The bond distances to ATP are essentially the same with or without enzyme, and the glycosidic bond angle of 90° is retained when the enzyme is present. These data indicate that Mn(II) is in fact strongly bound to the elongation site and that it holds nucleotide base in a manner suitable for complementary base pairing to the template, as would perhaps be expected at that site (219).

The distances from Mn(II) to the ApU at the initiation site in the presence of enzyme are much longer than the distances to ATP, and also much longer than in the binary Mn(II) ApU complex. These data indicate that Mn(II) does not bind directly to the initiator (219).

In the binary Mn(II) ApU complex the manganese appears to be intercalated between the adenine and uracil, that is, π bonded to these bases, while at the same time forming a σ bond to the phosphate group. The distances between Mn(II) and ApU on the enzyme indicate that the ApU has the bases in a skew orientation that is not favorable for hydrogen bonding to the DNA strand (Figure 32). It is postulated that perhaps the initiator is deliberately oriented in this way in order to prevent binding to DNA and to cause the release of the growing RNA strand from the DNA template as it passes through the initiator site.

Stein and Mildvan (227) were able to use a chromium complex of ATP (CrATP) as an initiator which did not function as an elongation substrate. At the same time they employed CTP as the elongation substrate, bound to the protein through Mn(II). They found no paramagnetic interaction between the Mn(II) and the Cr(III) by ESR and NMR techniques and concluded that the two metals were at least 10 to 11.5 Å apart. By calculating distances between Cr(III) and the ATP substrate, they determined that the substrate bases could stack with the initiator base and permit the transfer of the α phosphate group from the nucleotide on the elongation side to the nucleotide on the initiation site (227).

Metal ions can cause errors in RNA synthesis as in DNA synthesis. Substitution of Mn(II) for Mg(II) seems to improve fidelity with respect to recognition of complementary bases (220). On the other hand, Mn(II) induces the error incorporation of deoxynucleotides into RNA (221, 216); Mg(II) and Co(II) permit only the "correct" incorporation of ribonucleotides (216).

When RNA polymerase utilizes a DNA template bound to *cis*-[Pt(NH$_3$)$_2$Cl$_2$], the size of RNA molecules decreases as the amount of Pt bound to the DNA increases. Thus the Pt complex permits RNA synthesis but apparently prevents the enzyme from passing Pt binding sites on the template (222).

5.4 Metal Ions and Protein Synthesis

RNA molecules that have copied the DNA genes with the aid of RNA polymerase move into the cytoplasm as messenger RNA (mRNA). There they bind to particles called ribosomes on which the message is read. As the message moves past a ribosome, units of three bases, or codons, hydrogen bond to three complementary bases, or anticodons, on transfer RNA (tRNA) molecules which have amino acids attached at sites distant from the anticodon. Thus three consecutive bases on the message (mRNA) represent the code for one amino acid. Protein synthesis is concluded when peptide bonds are formed between the amino acids.

This very complicated process of translation involves metal ions in many stages. The ribosomes are composed of subunits, and Mg^{2+} ions affect the equilibrium between the ribosomes and their subunits (223). We have already noted the importance of metal ions in stabilizing the tRNA structure. Metal ions are required for the charging of tRNA with amino acids (224) and for the binding of tRNA to mRNA (225). We shall concern ourselves here briefly with one very intriguing phenomenon—the importance of metal ions in the fidelity of translation.

Metal ions (Mg^{2+}) are required for protein synthesis, but Szer and Ochoa (226) first demonstrated that an excess of Mg^{2+} can cause the incorporation of the wrong amino acid into protein. Using poly(U) as a message, they noted that the "correct" incorporation of phenylalanine into protein increased with increasing Mg^{2+} concentration to a maximum at 10 mM Mg^{2+} and decreased at higher Mg^{2+} concentration. However, at these higher Mg^{2+} concentrations leucine was misincorporated, maximally at approximately 20 mM Mg^{2+}.

A possible explanation for the misincorporation at high Mg^{2+} concentration could be the mispairing of bases in the codon–anticodon interaction. Such mispairing can be rationalized at high [Mg^{2+}], since high concentrations of metal ions stabilize polynucleotide strand interaction, and the codon–anticodon reaction could become so strong as to be no longer sensitive to the difference in stability of complementary base pairs and more weakly hydrogen-bonded base pairs. At low Mg^{2+} concentration the codon–anticodon interaction is relatively weak, and only the strongest H-bonded base pairs, namely, the complementary ones, are permitted.

This explanation would predict that low metal ion concentrations generally should favor only complementary base pairing and high metal ion concentrations should induce mispairing. This prediction was tested by reaction of a homopolymer [e.g., poly(A)] with a heteropolymer [e.g., poly(I,U)], so that complementary base pairing (A–U) and mispairing (A–I) could both occur. It was shown (222) that these synthetic polymers are in fact limited to complementary base pairing at low [Mg^{2+}] but permit mispairing at high [Mg^{2+}]. Curiously, the transition between complementary base pairing and mispairing occurs between 10 and 20 mM Mg^{2+}. Mispairing is therefore a plausible explanation for the misincorporation of amino acids into protein at high [Mg^{2+}]. Alternative explanations of the phenomenon are possible, however, because Mg^{2+} ions affect so many of the processes that are involved in protein synthesis.

ACKNOWLEDGMENT

The work described herein which was performed at The Johns Hopkins University was generously supported by NIHGMS Grant GM 20544.

REFERENCES

1. G. L. Eichhorn, in *Inorganic Biochemistry*, Vol. 2, G. L. Eichhorn, Ed., Elsevier, New York, 1973, p. 1191.

2. M. A. Sirover and L. A. Loeb, *Science*, **194**, 1434 (1976).

3. G. N. Schrauzer, Ed., *Inorganic and Nutritional Aspects of Cancer*, Vol. 91 of *Advances in Experimental Medicine and Biology*, Plenum Press, New York and London, 1977.

4. L. G. Marzilli, *Prog. Inorg. Chem.*, **23**, 255 (1977).

5. D. J. Hodgson, *Prog. Inorg. Chem.*, **23**, 211 (1977).

6. R. W. Gellert and R. Bau, in *Metal Ions in Biological Systems*, H. Sigel, Ed., Dekker, New York, 1979, p. 1

7. V. Swaminathan and M. Sundaralingam. *Crit. Rev. Biochem.*, in press.

8. R. B. Martin, in *Metal Ions in Biological Systems*, H. Sigel, Ed., Dekker, New York, 1979.

9. Ref. 1, p. 1216.

10. L. G. Marzilli and T. J. Kistenmacher, *Acc. Chem. Res.*, **10**, 146 (1977).

11. T. J. Kistenmacher and L. G. Marzilli, in *Metal–Ligand Interactions in Organic Chemistry and Biochemistry*, B. Pullman and N. Goldblum, Eds., D. Reidel, Dordrecht, Holland, 1977, Part 1, p. 7.

12. B. Rosenberg, *Naturwissenschaften*, **60**, 339 (1973).

13. M. L. De Pamphilis and W. W. Cleland, *Biochemistry*, **12**, 3714 (1973).

14. R. M. K. Dale and D. C. Ward, *Biochemistry*, **14**, 2458 (1975).

15. S. J. Lippard, *Acc. Chem. Res.*, **11**, 211 (1978).

16. C. Formoso, *Biochem. Biophys. Res. Commun.*, **53**, 1084 (1973).

17. *Acc. Chem. Res.*, **10** (1978).

18. C. M. Frey and J. Stuehr, *Met. Ions Biol. Sys.*, **1**, 51 (1973).

19. R. M. Izatt, J. J. Christiansen, and J. H. Rytting, *Chem. Rev.*, **71**, 439 (1971).

20. P. O. P. Ts'o, "Bases, Nucleosides, and Nucleotides," in *Basic Principles in Nucleic Acid Chemistry*, Vol. 1, P. O. P. Ts'o, Ed., Academic Press, New York–London, 1974, p. 453.

21. R. A. Newmark and C. R. Cantor, *J. Am. Chem. Soc.*, **90**, 5010 (1968).

22. T. L. V. Ulbricht, "Optical Rotatory Dispersion of Nucleosides and Nucleotides," in *Synthetic Procedures in Nucleic Acid Chemistry*, Vol. 2, W. W. Zorbach and R. S. Tipson, Eds., Wiley–Interscience, New York, 1973, p. 177.

23. M. Sundaralingam and J. A. Carrabine, *J. Mol. Biol.*, **61**, 287 (1971).

24. T. J. Kistenmacher, L. G. Marzilli, and C. H. Chang, *J. Am. Chem. Soc.*, **95**, 5817 (1973).

25. T. J. Kistenmacher, L. G. Marzilli, and D. J. Szalda, *Acta Crystallogr., Sect. B.*, **32**, 186 (1976).

26. P. deMeester, D. M. L. Goodgame, A. C. Skapski, and Z. Warnke, *Biochem. Biophys. Acta*, **324**, 301 (1973).

27. M. J. McCall and M. R. Taylor, private communication, 1974.

28. M. J. McCall and M. R. Taylor, *Biochem. Biophys. Acta*, **390**, 137 (1975).

29. C. J. L. Lock, R. A. Speramzini, G. Turner, and J. Powell, *J. Am. Chem. Soc.*, **98**, 7863 (1976).

30. C. Gagnon and A. L. Beauchamp, *Acta Crystallogr., Sect. B*, **33**, 1448 (1977).

31. M. R. Taylor, *Acta Crystallogr., Sect. B*, **29**, 884 (1973).

32. T. Sorrell, L. A. Epps, T. J. Kistenmacher, and L. G. Marzilli, *J. Am. Chem. Soc.*, **99**, 2173 (1977).

33. T. Sorrell, L. A. Epps, T. J. Kistenmacher, and L. G. Marzilli, *J. Am. Chem. Soc.*, **100**, 5756 (1978).

34. C. C. Chiang, L. A. Epps, L. G. Marzilli, and T. J. Kistenmacher, *Inorg. Chem.*, **18**, 791 (1979).

35. T. J. Kistenmacher, T. Sorrell, M. Rossi, C. C. Chiang, and L. G. Marzilli, *Inorg. Chem.*, **17**, 479 (1978).

36. A. D. Collins, P. deMeester, D. M. L. Goodgame, and A. C. Skapski, *Biochim. Biophys. Acta*, **402**, 1 (1975).

37. E. A. Merritt, M. Sundaralingam, R. D. Cornelius, and W. W. Cleland, *Biochemistry*, in press.

38. L. G. Marzilli, R. C. Stewart, C. P. Van Vuuren, B. de Castro, and J. P. Caradonna, *J. Am. Chem. Soc.*, **100**, 3967 (1978).

39. S. Mansy, T. E. Wood, J. C. Sprowles, and R. S. Tobias, *J. Am. Chem. Soc.*, **96**, 1762 (1974).

40. G. Y. H. Chu, R. E. Duncan, and R. S. Tobias, *Inorg. Chem.*, **16**, 2625 (1977).

41. R. W. Christian, S. Mansy, H. J. Peresie, A. Ranade, T. A. Berg, and R. S. Tobias, *Bioinorg. Chem.*, **7**, 245 (1977).

42. G. Y. H. Chu and R. S. Tobias, *J. Am. Chem. Soc.*, **98**, 2641 (1976).

43. S. Mansy and R. S. Tobias, *Biochemistry*, **14**, 2952 (1975).

44. A. T. Tu and M. J. Heller, *Met. Ions Biol. Syst.*, **1**, 1 (1973).

45. T. J. Kistenmacher, D. J. Szalda, and L. G. Marzilli, *Inorg. Chem.*, **14**, 1686 (1975).

46. N. A. Berger and G. L. Eichhorn, *Biochemistry*, **10**, 1847 (1971).

47. G. L. Eichhorn, P. Clark, and E. D. Becker, *Biochemistry*, **5**, 245 (1966).

48. L. G. Marzilli, W. C. Trogler, D. P. Hollis, T. J. Kistenmacher, C. H. Chang, and B. E. Hanson, *Inorg. Chem.*, **14**, 2568 (1975).

49. P. C. Kong and T. Theophanides, *Inorg. Chem.*, **13**, 1981 (1974).

50. R. J. Angelici, in *Inorganic Biochemistry*, Vol. 1, G. L. Eichhorn, ed., Elsevier, New York, 1973, p. 63.

51. H. Sigel, *J. Inorg. Nucl. Chem.*, **39**, 1903 (1977).

52. T. A. Glassman, C. Cooper, L. W. Harrison, and T. J. Swift, *Biochemistry*, **10**, 843 (1971).

53. H. Sternglanz, E. Subramanian, J. C. Lacey, and C. E. Bugg, *Biochemistry*, **15**, 4797 (1976).

54. M. T. Cohn and T. R. Hughes, Jr., *J. Biol. Chem.*, **237**, 176 (1962).

55. Tran-Dinh Son, M. Roux, and M. Ellenberger, *Nucleic Acids Res.*, **2**, 1101 (1975).

56. S. Tran-Dinh and J. M. Neumann, *Nucleic Acids Res.*, **4**, 397 (1977).

57. R. B. Martin. *Abstr. 12th MARM. ACS*, 1978, Abstr. Inorg. 36. Also see section 10.1 of Ref. 8.

58. L. G. Marzilli, B. de Castro, J. P. Caradonna, R. C. Stewart and C. P. Van Vuuren, *J. Am. Chem. Soc.*, in press.

59. W. W. Cleland and A. S. Mildvan, in *Advances in Inorganic Biochemistry*, Vol. 1, G. L. Eichhorn and L. G. Marzilli, Eds., Elsevier, Amsterdam, 1979, p. 163.

60. R. D. Cornelius, P. A. Hart, and W. W. Cleland, *Inorg. Chem.*, **16**, 2799 (1977).

61. D. G. Gorenstein, *J. Am. Chem. Soc.*, **97**, 898 (1975).

62. L. Rimai and M. E. Heyde, *Biochem. Biophys. Res. Commun.*, **41**, 313 (1970).

63. L. Rimai, M. E. Heyde, and E. B. Caren, *Biochem. Biophys. Res. Commun.*, **38**, 231 (1970).

64. M. E. Heyde and L. Rimai, *Biochemistry*, **10**, 1121 (1970).

65. J. F. Conn, J. J. Kim, F. L. Suddath, P. Blattman, and A. Rich, *J. Am. Chem. Soc.*, **96**, 7152 (1974).

66. K. Aoki, *Acta Crystallogr., Sect. B*, **32**, 1454 (1976).

67. R. W. Gellert and R. Bau, *J. Am. Chem. Soc.*, **97**, 7379 (1975).

68. (a) K. Aoki, G. R. Clark, and J. D. Orbell, *Biochim. Biophys. Acta*, **425**, 369 (1976); (b) E. Sletten and B. Lie, *Acta Crystallogr., Sect. B*, **32**, 3301 (1976).

69. S. T. Rao and M. Sundaralingam, *J. Chem. Soc., Chem. Commun.*, 995 (1968).

70. G. R. Clark and J. D. Orbell, *J. Chem. Soc., Chem. Commun.*, 139 (1974).

71. D. M. L. Goodgame, I. Jeeves, F. L. Phillips, and A. C. Skapski, *Biochim. Biophys. Acta*, **378**, 153 (1975).

72. R. Bau, R. W. Gellert, S. M. Lehovec, and S. Louie, *J. Clin. Hematol. Oncol.*, **7**, 51 (1977).

73. R. E. Cramer and P. L. Dahlstrom, *J. Clin. Hematol. Oncol.*, **7**, 330 (1977).

74. T. Sorrell, L. G. Marzilli, and T. J. Kistenmacher, *J. Am. Chem. Soc.*, **98**, 2181 (1976).

75. D. J. Szalda, T. J. Kistenmacher, and L. G. Marzilli, *Inorg. Chem.*, **15**, 2783 (1976).

76. D. J. Szalda, T. J. Kistenmacher, and L. G. Marzilli, *J. Am. Chem. Soc.*, **98**, 8371 (1976).

77. R. B. Simpson, *J. Am. Chem. Soc.*, **86**, 2059 (1964).

78. G. L. Eichhorn, P. Clark, and E. D. Becker, *Biochemistry*, **5**, 245 (1966).

79. N. Hadjiliadis and T. Theophanides, *Inorg. Chim. Acta*, **16**, 77 (1976).

80. J. Dehand and J. Jordanov, *J.C.S. Chem. Commun.*, 598 (1976).

81. G. Pneumatikakis, N. Hadjiliadis, and T. Theophanides, *Inorg. Chim. Acta*, **22**, L1 (1977).

82. D. M. L. Goodgame, I. Jeeves, F. L. Phillips, and A. C. Skapski, *Biochim. Biophys. Acta*, **378**, 153 (1975).

83. H. I. Heitner and S. J. Lippard, *Inorg. Chem.*, **13**, 815 (1974).

84. E. Sletten and A. Apeland, *Acta Crystallogr., Sect. B*, **31**, 2019 (1975).

85. (a) M. R. Caira and L. R. Nassimbeni, *Acta Crystallogr., Sect. B*, **31**, 1339 (1975); (b) A. L. Shoemaker, P. Singh, and D. J. Hodgson, *Acta Crystallogr., Sect. B*, **32**, 979 (1976); (c) L. Pope, M. Liang, M. R. Caira, and L. R. Nassimbeni, *Acta Crystallogr., Sect. B*, **32**, 612 (1976).

86. P. Lavertue, J. Hubert, and A. L. Beauchamp, *Inorg. Chem.*, **15**, 322 (1976).

87. D. M. L. Goodgame, I. Jeeves, G. D. Reynolds, and A. C. Skapski, *Nucleic Acids Res.*, **2**, 1375 (1975).

88. H. J. Krentzien, M. J. Clark, and H. Taube, *Bioinorg. Chem.*, **4**, 143 (1975).

89. M. Sundaralingam and J. A. Carrabine, *J. Mol. Biol.*, **61**, 287 (1971).

90. D. J. Szalda, L. G. Marzilli, and T. J. Kistenmacher, *Biochem. Biophys. Res. Commun.*, **63**, 601 (1975).

91. M. Authier-Martin and A. L. Beauchamp, *Can. J. Chem.*, **55**, 1213 (1977).

92. R. Melanson and F. D. Rochon, *Inorg. Chem.*, **17**, 679 (1978).

93. C. J. L. Lock, R. A. Speranzini, and J. Powell, *Can. J. Chem.*, **54**, 53 (1976).

94. E. Sinn, C. M. Flynn, and R. Bruce Martin, *Inorg. Chem.*, **16**, 2403 (1977).

95. S. Louie and R. Bau, *J. Am. Chem. Soc.*, **99**, 3874 (1977).

96. K. Aoki, *J. Chem. Soc., Chem. Commun.*, 748 (1976).

97. L. G. Marzilli, T. J. Kistenmacher, and M. Rossi, *J. Am. Chem. Soc.*, **99**, 2797 (1977).

98. M. J. Heller, A. J. Jones, and A. T. Tu, *Biochemistry*, **9**, 4981 (1970).

99. R. C. Lord and G. J. Thomas, Jr., *Spectrochim. Acta, Part A*, **23**, 2551 (1967).

100. K. W. Jennette, S. J. Lippard, and D. A. Ucko, *Biochem. Biophys. Acta*, **402**, 403 (1975).

101. D. J. Nelson, P. L. Yeagle, T. L. Miller, and R. B. Martin, *Bioinorg. Chem.*, **5**, 353 (1976).

102. J. Dehand and J. Jordanov, *J. Chem. Soc., Dalt.*, 1588 (1977).

103. D. Perahia, A. Pullman and B. Pullman, *Theor. Chim. Acta* (Berlin), **43**, 207 (1977).

104. See Ref. 48.

105. G. Kotowycz, *Can. J. Chem.*, **52**, 924 (1974).

106. T. Yokono, S. Shimokawa, and J. Sohma, *J. Am. Chem. Soc.*, **97**, 3827 (1975).

107. L. G. Marzilli, J. P. Caradonna, and B. de Castro, unpublished results.

108. R. Faggiani, B. Lippert, and C. J. L. Lock, *J. Am. Chem. Soc.*, submitted.

109. J. A. Carrabine and M. Sundaralingam, *Biochemistry*, **10**, 292 (1971).

110. B. E. Fischer and R. Bau, *J. Chem. Soc., Chem. Commun.*, 272 (1977), *Inorg. Chem.*, **17**, 27 (1978).

111. L. D. Kosturko, C. Folzer, and R. F. Stewart, *Biochemistry*, **13**, 3949 (1974).

112. T. J. Kistenmacher, L. G. Marzilli, and M. Rossi, *Bioinorg. Chem.*, **6**, 347 (1976).

113. M. C. Lim and R. B. Martin, *J. Inorg. Nucl. Chem.*, **38**, 1915 (1976).

114. N. A. Berger, E. Tarien, and G. L. Eichhorn, *Nature, New Biol.*, **239**, 237 (1972).

115. G. Brun, D. M. L. Goodgame, and A. C. Skapski, *Nature*, **253**, 127 (1975).

116. S. J. Kirchner, Q. Fernando, and M. Chvapil, *Inorg. Chim. Acta*, **25**, 245 (1977).

117. P. Chalilpoyil, and L. G. Marzilli, *Inorg. Chem.*, **18**, 2328 (1979).

118. Y. H. Chao and D. R. Kearns, *J. Am. Chem. Soc.*, **99**, 6425 (1977).

119. F. B. Daniel and E. J. Behrman, *J. Am. Chem. Soc.*, **97**, 7352 (1975).

120. J. P. Davidson, P. J. Faber, R. G. Fischer, S. Mansy, H. J. Peresie, B. Rosenberg, and L. Van Camp, *Chemother. Rep.*, **59**, 287 (1975).

121. J. K. Barton, H. N. Rabinowitz, D. J. Szalda, and S. J. Lippard, *J. Am. Chem. Soc.*, **99**, 2827 (1977).

122. B. K. Teo, K. Kijima and R. Bau, *J. Am. Chem. Soc.*, **100**, 621 (1978).

123. D. Voet and A. Rich, *Prog. Nucleic Acid Res. Mol. Biol.*, **10**, 183 (1970).

124. M. Rossi and T. J. Kistenmacher, *Acta Crystallogr., Sect. B*, **33**, 3962 (1977).

125. M. Rossi and T. J. Kistenmacher, to be published.

126. E. Shefter, S. Singh, T. Brennan, and P. Sackman, *Cryst. Struct. Commun.*, **3**, 209 (1976).

127. D. J. Szalda, L. G. Marzilli, and T. J. Kistenmacher, *Inorg. Chem.*, **14**, 2076 (1975).

128. T. J. Kistenmacher, D. J. Szalda, and L. G. Marzilli, *Acta Crystallogr., Sect. B*, **31**, 2416 (1975).

129. G. R. Clark and J. D. Orbell, *J. Chem. Soc., Chem. Commun.*, 697 (1975).

130. D. J. Szalda, T. J. Kistenmacher, and L. G. Marzilli, *Inorg. Chem.*, **14**, 2623 (1975).

131. E. Sletten and B. Thorstensen, *Acta Crystallogr., Sect. B*, **30**, 2483 (1974).

132. E. Sletten and M. Ruud, *Acta Crystallogr., Sect. B*, **31**, 982 (1975).

133. T. J. Kistenmacher, D. J. Szalda, C. C. Chiang, M. Rossi, and L. G. Marzilli, *Inorg. Chem.*, **17**, 2582 (1978).

134. M. Authier-Martin, J. H. Hubert, R. Rivest, and A. L. Beauchamp, *Acta Crystallogr., Sect. B*, **34**, 273 (1978).

135. R. F. Whiting and F. P. Ottensmeyer, *Biochim. Biophys. Acta*, **474**, 334 (1977).

136. C. H. Chang, M. Beer, and L. G. Marzilli, *Biochemistry*, **16**, 33 (1977).

137. L. R. Subbaraman, J. Subbaraman, and E. J. Behrman, *Bioinorg. Chem.*, **1**, 35 (1971).

138. N. A. Berger and G. L. Eichhorn, *Biochemistry*, **10**, 1857 (1971).

139. A. Yamada, K. Akasaka, and H. Hatano, *Biopolymers*, **15**, 1315 (1976).

140. A. Jack, J. E. Ladner, D. Rhodes, R. S. Brown, and A. Klug, *J. Mol. Biol.*, **111**, (1977).

141. A. Jack, J. E. Ladner, and A. Klug. *J. Mol. Biol.*, **108**, 619 (1976).

142. L. L. Muchausen and R. O. Rahn, *Biochim. Biophys. Acta*, **414**, 242 (1975).

143. C. C. Chiang, T. Sorrell, T. J. Kistenmacher, and L. G. Marzilli, *J. Am. Chem. Soc.*, **100**, 5102 (1978).

144. S. J. Lippard, P. J. Bond, K. C. Wu, and W. R. Bauer, *Science*, **194**, 726 (1976).

145. S. K. Arya and J. T. Yang, *Biopolymers*, **14**, 1847 (1975).

146. N. Davidson, J. Widholm, U. S. Nandi, R. Jensen, B. M. Olivera, and J. C. Wang, *Proc. Natl. Acad. Sci. USA*, **53**, 111 (1965).

147. S. Arnot and D. W. L. Hukins, *J. Mol. Biol.*, **81**, 93 (1973).

148. N. A. Berger and G. L. Eichhorn, *J. Am. Chem. Soc.*, **93**, 7062 (1971).

149. A. Yamada, K. Akasaka, and H. Hatano, *Biopolymers*, **15**, 1315 (1976).

150. C. Zimmer, G. Luck, H. Fritzsche, and H. Triebel, *Biopolymers*, **10**, 441 (1971).

151. R. H. Jensen and N. Davidson, *Biopolymers*, **4**, 17 (1966).

152. P. J. Stone, A. D. Kelman, and F. M. Sinex, *Nature*, **251**, 736 (1974).

153. T. Yamane and N. Davidson, *J. Am. Chem. Soc.*, **83**, 2599 (1961).

154. N. Davidson, J. Widholm, U. S. Nandi, R. Jensen, B. M. Olivera, and J. C. Wang, *Proc. Natl. Acad. Sci. USA*, **53**, 111 (1965).

155. G. Yonuschot and G. W. Mushrush, *Biochemistry*, **14**, 1677 (1975).

156. Y. A. Shin, J. M. Heim, and G. L. Eichhorn, *Bioinorg. Chem.*, **1**, 149 (1972).

157. Y. A. Shin, *Biopolymers*, **12**, 2459 (1973).

158. Y. A. Shin, unpublished results.

159. Y. A. Shin and G. L. Eichhorn, *Biochemistry*, **7**, 1026 (1968).

159a. J. M. Rifkind and G. L. Eichhorn, *J. Am. Chem. Soc.*, **94**, 6526 (1972).

160. Y. A. Shin and G. L. Eichhorn, *Biochemistry*, **7**, 1026 (1968).

161. K. W. Jennette, S. J. Lippard, G. A. Vassiliades, and W. R. Bauer, *Proc. Natl. Acad. Sci. USA*, **71**, 3839 (1974).

162. J. M. Rifkind, Y. A. Skin, J. M. Heim, and G. L. Eichhorn, *Biopolymers*, **15**, 1879 (1976).

163. H. Krakaver, *Biopolymers*, **10**, 2459 (1971).

164. G. L. Eichhorn and P. Clark, *Proc. Natl. Acad. Sci. USA*, **53**, 586 (1965).

165. G. L. Eichhorn and Y. A. Shin, *J. Am. Chem. Soc.*, **90**, 7323 (1968).

166. S. Hiai, *J. Mol. Biol.*, **11**, 672 (1965).

167. H. Venner and C. Zimmer, *Biopolymers*, **4**, 321 (1966).

168. J. H. Coates, D. O. Jordan, and V. K. Srivastava, *Biochem. Biophys. Res. Commun.*, **20**, 611 (1965).

169. T. Yamane and N. Davidson, *J. Am. Chem. Soc.*, **83**, 2599 (1961).

170. D. W. Gruenwedel and N. Davidson, *J. Mol. Biol.*, **21**, 129 (1966).

171. D. W. Gruenwedel and N. Davidson, *Biopolymers*, **5**, 847 (1967).

172. G. L. Eichhorn, J. J. Butzow, P. Clark, and E. Tarien, *Biopolymers*, **5**, 283 (1967).

173. T. Lindahl, A. Adams, and J. R. Fresco, *Proc. Natl. Acad. Sci. USA*, **55**, 941 (1966).

173a. R. L. Karpel, N. S. Miller, A. M. Lesk, and J. R. Fresco, *J. Mol. Biol.*, **97**, 519 (1975).

174. M. Cohn, A. Danchin, and M. Grunberg-Manago, *J. Mol. Biol.*, **39**, 199 (1969).

175. R. Römer and R. Hach, *Eur. J. Biochem.*, **55**, 271 (1975).

176. A. Danchin, *Biopolymers*, **11**, 1317 (1972).

177. M. S. Kayne and M. Cohn, *Biochemistry*, **13**, 4159 (1974).

178. P. H. Bolton and D. R. Kearns, *Biochemistry*, **16**, 5729 (1977).

179. A. H. Schrier and P. R. Schimmel, *J. Mol. Biol.*, **93**, 323 (1975).

180. A. Stein and D. M. Crothers. *Biochemistry*, **15**, 157 (1976).

181. S. R. Holbrook, J. L. Sussman, R. W. Warrant, G. M. Church, and S.-H. Kim, *Nucleic Acids Res.*, **4**, 2811 (1977).

182. A. Jack, J. E. Ladner, D. Rhodes, R. S. Brown, and A. Klug, *J. Mol. Biol.*, **111**, 315 (1977).

183. D. Rhodes, P. W. Piper, and B. F. C. Clark, *J. Mol. Biol.*, **89**, 469 (1974).

184. H. R. Sunshine and S. J. Lippard, *Nucleic Acids Res.*, **1**, 673 (1974).

185. K. Dimroth, L. Jaenicke, and D. Heinzel, *Ann. Chem.*, **566**, 206 (1950).

186. E. Bamann, H. Trapmann, and F. Fischler, *Biochem. Z.*, **328**, 89 (1954).

187. G. L. Eichhorn and J. J. Butzow, *Biopolymers*, **3**, 79 (1965).

188. J. J. Butzow and G. L. Eichhorn, *Biopolymers*, **3**, 95 (1965).

189. W. R. Farkas, *Biochim. Biophys. Acta*, **155**, 401 (1968).

190. J. J. Butzow and G. L. Eichhorn, *Biochemistry*, **10**, 2019 (1971).

191. J. J. Butzow and G. L. Eichhorn, *Nature*, **254**, 358 (1975).

192. G. L. Eichhorn, E. Tarien, and J. J. Butzow, *Biochemistry*, **10**, 2014 (1971).

193. C. Werner, B. Krebs, G. Keith, and G. Dirheimer, *Biochim. Biophys. Acta*, **432**, 161 (1976).

194. H. Sawai and L. E. Orgel, *J. Am. Chem. Soc.*, **97**, 3532 (1975).

195. H. Sawai, *J. Am. Chem. Soc.*, **98**, 7037 (1976).

196. P. Clark and G. L. Eichhorn, *Biochemistry*, **13**, 5098 (1974).

197. Y. A. Shin and G. L. Eichhorn, *Biopolymers*, **16**, 225 (1977).

198. A. Monneron and Y. Moule, *Exp. Cell Res.*, **51**, 531 (1968).

199. Y-J. Kang, M. O. J. Olson, and H. Busch, *J. Biol. Chem.*, **249**, 5580 (1974).

200. M. S. Kanungo and M. K. Thakur, *Biochem. Biophys. Res. Commun.*, **79**, 1031 (1977).

201. J. P. Slater, A. S. Mildvan, and L. A. Loeb, *Biochem. Biophys. Res. Commun.*, **44**, 37 (1971).

202. D. L. Sloan, L. A. Loeb, A. S. Mildvan, and R. J. Feldmann, *J. Biol. Chem.*, **250**, 8913 (1975).

203. M. Sundaralingam, *Biopolymers*, **7**, 821 (1969).

204. M. A. Sirover and L. A. Loeb, *Biochem. Biophys. Res. Commun.*, **70**, 812 (1976).

205. M. A. Sirover and L. A. Loeb, *J. Biol. Chem.*, **252**, 3605 (1977).

206. D. K. Dube and L. A. Loeb, *Biochem. Biophys. Res. Commun.*, **67**, 1041 (1975).

207. M. A. Sirover and L. A. Loeb, *Proc. Natl. Acad. Sci. USA.* **73**, 2331 (1976).

208. M. A. Sirover and L. A. Loeb, *Science*, **194**, 1434 (1976).

209. P. Berg, H. Fancher, and M. Chamberlain, in *Symposium on Informational Macromolecules*, H. Vogel, Ed., Academic Press, New York, 1963, p. 467.

210. W. Salser, K. Fry, C. Brunk, and R. Poon, *Proc. Natl. Acad. Sci. USA*, **69**, 238 (1972).

211. W. A. Salser, *Ann. Rev. Biochem.*, **43**, 923 (1974).

212. M. C. Scrutton, C. W. Wu, and D. A. Goldthwait, *Proc. Natl. Acad. Sci. USA*, **68**, 2497 (1971).

213. D. C. Speckhard, F. Y.-H. Wu, and C. W. Wu, *Biochemistry*, **16**, 5228 (1977).

214. P. Ballard and H. G. Williams-Ashman, *Nature*, **203**, 150 (1964).

215. C. C. Widnell and J. R. Tata, *Biochem. J.*, **92**, 313 (1964).

216. K. G. Rao, F. Y.-H. Wu, S. Sethi, and G. L. Eichhorn, Abstract, 174th National Meeting of ACS, Aug.–Sept. 1977.

217. K. M. Downey and A. G. So, *Biochemistry*, **9**, 2520 (1970).

218. R. Koren and A. S. Mildvan, *Biochemistry*, **16**, 241 (1977).

219. B. L. Bean, R. Koren, and A. S. Mildvan, *Biochemistry*, **16**, 3322 (1977).

220. C. F. Springgate and L. A. Loeb, *J. Mol. Biol.*, **97**, 577 (1975).

221. J. Hurwitz, L. Yarbrough, and S. Wickner, *Biochem. Biophys. Res. Commun.*, **48**, 628 (1972).

222. G. L. Eichhorn, J. Rifkind, Y. A. Shin, J. Pitha, J. Butzow, P. Clark, and J. Froehlich, in *Metal–Ligand Interactions in Organic Chemistry and Biochemistry*, B. Pullman and N. Goldblum, Eds., D. Reidel Publ., Dordrecht, Holland, 1977, Part 1, pp. 41–51.

223. A. Goldberg, *J. Mol. Biol.*, **15**, 663 (1966).

224. J. E. Allende, G. Mora, M. Gatica, and C. C. Allende, *J. Biol. Chem.*, **240**, PC3229 (1965).

225. T. Ohta, S. Sarkar, and R. E. Thach, *Proc. Natl. Acad. Sci. USA*, **58**, 1638 (1967).

226. W. Szer and S. Ochoa, *J. Mol. Biol.*, **8**, 823 (1964).

227. P. J. Stein and A. S. Mildvan, *Biochemistry*, **17**, 2675 (1978).

Index

Acceptor stem region of tRNA, 164-166
Activation by metal ions, effect on fidelity, 130-132
Adenosine, binding by osmium, 44
pK$_a$'s, 38-39
structure, 34-35
see also Nucleosides
Adenosine derivatives, metal binding to, 189-198
metal-phosphate bonding, 191
Altered enzyme conformation, 139
Altered substrate conformation, 139
Altered template base specificity, 139-140
AMP, complex with platinum terpyridine chloride, 55
AMV, 126, 130, 131, 132, 139
Analysis of metals by alterations in fidelity of DNA synthesis, 133-134
Animal cancers, activity of platinum complexes against, 16-17
Anticancer activity of platinum complexes, 10-15
Anticancer drugs, molecular structure of, 15-18
uncertainty over function of, 20-21
Anticodon stem and loop of tRNA, 155-158
Anticodons, 242
Antitumor drug, platinum, 88-106
see also cis-DDP

Bacterial cells, effects on of platinum complexes, 6-10
metal mutagenesis in, 118-119

treatment of with electric fields, 3-5
Base, interactive, 44, 46-47
metal binding, common sites, 46-47
through exocyclic substituents, 49-55
in polynucleotides, 67-71
weak interactions, 54-55
numbering scheme, 34-35
protonation sites, 38-39
Base pairs, 40-41
in [(terpy)PtC1](5'-AMP), 55
Base specificity, 47
of *cis*-DDP, 99
of mercury, 85-86
of platinum intercalators, 75
of silver, 85-86
Bifurcated hydrogen bonding, 196
Binding modes, types of, 186-189
Binding sites, 186-189
crystallographic interpretation of, 151
Biological processes, role of metal ions in, 236-243
Buoyant density of DNA, 60
effects of metal binding, 85, 99
as purification technique, 85-86

Cadmium, binding to DNA, 67-68
binding to IMP, 58, 59
Cancer, treatment of with platinum complexes, 3-29
Cancer chemotherapy, uncertainty over operation of, 20-21
Cancer destruction, explanation of, 21-25
Carcinogenesis, in mammals, 120-121
relationship to mutagenesis, 117-118